Bacteriophages - Perspectives and Future

Edited by Renos Savva

Published in London, United Kingdom

IntechOpen

Supporting open minds since 2005

Bacteriophages - Perspectives and Future
http://dx.doi.org/10.5772/intechopen.73439
Edited by Renos Savva

Contributors
James Zahn, Mathew Halter, Wibool Piyawattanametha, Supang Khondee, Elena Orlova, Helen White, John Ward, Vitor Pinheiro, Paul Rothwell, Sophie Knott, Sarah A Milsom, Renos Savva

First published in London, United Kingdom, 2020 by IntechOpen
IntechOpen is the global imprint of INTECHOPEN LIMITED, registered in England and Wales, registration number: 11086078, 7th floor, 10 Lower Thames Street, London, EC3R 6AF, United Kingdom
Printed in Croatia

British Library Cataloguing-in-Publication Data
A catalogue record for this book is available from the British Library

Additional hard and PDF copies can be obtained from orders@intechopen.com

Bacteriophages - Perspectives and Future
Edited by Renos Savva
p. cm.
Print ISBN 978-1-83880-438-1
Online ISBN 978-1-83880-446-6
eBook (PDF) ISBN 978-1-83880-447-3

For EU product safety concerns:
IN TECH d.o.o., Prolaz Marije Krucifikse Kozulić 3, 51000 Rijeka, Croatia, info@intechopen.com or visit our website at intechopen.com.

Meet the editor

Renos Savva is a senior lecturer at Birkbeck, University of London. His research interests encompass the biological interactions of virus proteins, both with nucleic acids and with other proteins. Dr Savva's research concerns the nature of viral survival mechanisms, such as evasion of host restriction factors, and the adaptation and sequence plasticity of virus-encoded proteins. Dr Savva's published research includes insights from structural biology of phage-encoded inhibitors of the ubiquitous Ung enzymes. His research furthermore addresses questions of protein sequence plasticity, which crosses over into synthetic biology collaborations: How adaptable are essential phage proteins, and what can be achieved in molecular engineering of phage particles and payloads? He also advocates for commercial translation of such fundamental insights for societal benefit.

Contents

Preface

Bacteriophages are viruses that target and infect bacterial cells. As such, they have long attracted interest as potential therapeutic agents to control microbial pathogenesis. Since their discovery they have been proven to be promising antimicrobials, but caution has been mixed into that hope, because they are also capable of invoking pathogenicity in commensal and environmental species of bacteria. Study and cumulative knowledge of the ingenious factors that these viruses use to survive in the wild and to prevail over their host cells, has contributed greatly to many diverse facets of molecular biology and to the rise of synthetic biology.

There is once again great promise that bacteriophages will play a prominent role in healthcare, not least fuelled by the rise in microbial antibiotic resistance globally. However, this hope also presents challenges. To understand the challenges, it is as well to also regard the problems bacteriophages could pose and how these can be solved. To realise their applied promise, bacteriophages will require production at unprecedented industrial scales. Exploring efficiency of mass production, as well as how to ensure purity from other bacteriophages that could contaminate medical-grade stocks, will be essential. In this volume, the chapters contributed respectively, by authors Zahn and Halter, and Ward et al., bring these matters into focus.

In a wider healthcare context, the structural properties of bacteriophages become important factors for consideration. The chapter in this volume contributed by White and Orlova gives a detailed overview of what we know so far about the structure and mechanics of some of the better-known phage particle types. Engineering the surfaces of these viruses is a mature technology, but its deployment in personalised medical imaging is now technologically feasible and demonstrated as effective. The contributed chapter by Khondee and Piyawattanametha illuminates how surgery to remove metastatic tumours is given greater precision via engineered phage particle avidity. Exploiting the unusual properties of bacteriophage-encoded enzymes can allow new approaches in gene therapies and cell and tissue engineering. In fact, applications enabled by such unusual enzymatic activities could revolutionise data storage via biological macromolecules. The chapters contributed to this volume by Pinheiro, and the chapter by Knott, Milsom, and Rothwell, showcase unusual enzymatic processes which bacteriophages utilise for their adaptation and survival.

Taking all these themes together, it is possible to envision bacteriophages as potentially versatile nanoscale entities, with scope to be repurposed for a variety of interventions involving bacterial cells, or in entirely unrelated aims, encompassing aspects of healthcare and electronics. The macromolecular tools utilised by replicating bacteriophages can further enrich such endeavours.

This book presents facets of current knowledge relevant to realising those broader future aims.

Renos Savva
Institute of Structural and Molecular Biology,
Department of Biological Sciences, Birkbeck,
University of London, United Kingdom

Chapter 1

Introductory Chapter: Nature's Ancient Nanomachines and Their Synthetic Future

Renos Savva

1. Resurgence

Bacteriophages are generally considered to be the most prevalent biological entities on planet Earth [1], with astronomical estimates of their myriad abundance. These fascinatingly diverse viruses, which infect bacterial cells, were discovered in the second decade of the twentieth century. Under the light microscope, bacterial cells were seen to be apparently eaten away, hence the scientific Greek naming of these viruses meaning literally "bacteria eaters". The middle decades of that century gave us a first glimpse of these viral particles, via their imaging using the technique of negative stain electron microscopy. The morphology of the first studied particles was symbolic of the so-called Space Age in which these discoveries took place: The archetypal Phi X 174 evoked the first artificial satellites, while the T-even phages resembled prototypic lunar landing vehicles.

Such latter-day futuristic symbolism may seem amusingly outdated by the time of writing. However, the undoubtedly ancient bacteriophages are now riding a new wave of technological advancement: synthetic biology. Interestingly, this returns us to the potential revolution in healthcare that surrounded the initial discovery of bacteriophages. The simple fact is that *phages*, which is a common abbreviation used when referring to these viruses, are capable of selective killing of bacterial cells of a given species or strain. In other words, phages are exquisitely specific microbial control agents. In the current antimicrobial resistance era [2], exemplified by the ESKAPE organisms, which is to name only the vanguard of untreatable pathogenic microbes, phages offer a potential panacea for the treatment of pathogenic bacterial infection.

Indeed, deployment of phages to treat bacterial infections in animals and humans was the first exploitative use of these virus entities. Their promise as therapeutic agents was, nevertheless, soon eclipsed by the rise of antibiotics. However, partly by economic necessity, the refinement of such phage-mediated treatments persisted in the former USSR [3]. Today, the fruit of that legacy provides effective alternative treatment in an era of antibiotic-resistant microorganisms [4]. Including the long-standing cases of treatment achieved within the pioneering hospitals in the former Soviet territory of Georgia [5], there have also been recent high-profile cases [6], as well as a rise in commercial offerings based upon sound scientific discovery [7].

2. Repurposed

In a commercial sense, phages are regarded as cheap to manufacture, because they will naturally multiply within their target bacteria. However, process control at

IntechOpen

industrial scales presents challenges: most notably in efficiency and reproducibility but also via complications in terms of sterility and purity [7, 8]. Knowing how to keep prevalent diverse phage types out of a production facility will doubtless be as important as manufacturing commercially relevant types efficiently. When considering efficiency of phage production, the ultimate cost depends upon the measures that have to be taken, from plant and tooling to culture volume and starter culture characteristics. Investigations of multiplicity of infection, and economics of scale, plus processing of waste effluent are all factors that will impact costs and affordability [9]. Upscale and purity of phage will have an impact on any type of technology or application that is currently envisaged for phage, from antimicrobial therapy to phage particles as nanomaterials.

We can consider that phage genetic and structural insights are opening doors in nanotechnology and synthetic biology applications, which have a translation potential back into healthcare. That medical relevance is not limited to antimicrobial applications but also encompasses cancer and gene therapy. Indeed, phage display technologies are revolutionising the high-resolution visualisation of metastatic tumours in surgical settings by enabling unambiguous contrast.

Regarding the structure of phage particles, as nanoscale parts assembling into a mechanised vehicle for DNA packaging and delivery, their production as potential therapeutic agents might well be due a synthetic makeover. The concept of a phage cocktail is now very well tested: namely, several phage types recognising the same bacterial species but attacking that cell type in different ways. Phage cocktails are known to make it less likely that the targeted cells can survive via adaptation, as is known to be the case via treatment with a single phage type. Lately, the reprogramming of a phage DNA payload to ensure target cell death rather than phage latent persistence in a population of cells has also been found to be effective [6]. Indeed, such engineering approaches have been used to alter the DNA payload injected by the phage particle in order to make phages more effective at cell killing, rather than infection per se [6, 10].

Nevertheless, from an efficiency and medical regulatory perspective, it could be advantageous to have one or just a few medically approved phage chassis. These might be envisaged as phages known for their low immunogenicity profiles in patients. Perhaps these may be patient-specific, in which the therapeutic DNA payload is installed in a phage chassis matched to the current patient's immune tolerance. Then one can envisage a programmable targeting built in via a synthetic biology approach of re-engineering components of these standardised phages to recognise any desired bacterial target. Demonstrations of swapping out the targeting structures (i.e., tails and adsorption features) have been equally impressive in principle [11, 12]. Thus, furthering a detailed knowledge of phage structure beyond well-studied phage types [13] is incentivised.

3. Reimagined

One of the most fascinating things about phages is that they are hotbeds of molecular adaptation. An eclectic and presently largely arcane phage-encoded protein panoply supports the survival and success of the diverse bacteriophage types. The exploitation of deciphered elements of this phage protein repertoire was central to development of the recombinant biotechnology revolution. Circuitously, these very molecular tools are bringing about the technological revolution that promises to lend a starring role of phage in a biotechnologically repurposed guise. In fact, looking into that mesmerising pool of proteins of unknown function [14], it is easy to believe that the phages' encoded box of tricks isn't yet done revolutionising applied biotechnology. The numbers of completely sequenced diverse phage genomes expand at a prolific rate, and the mysteries of their encoded protein manifest may yet take a long time and a lot of care to unravel.

Phage biology then, played a central role in the elucidation of some of the major genetic insights of the twentieth century. Other ground-breaking studies in that century concerned phage genome replication, the structures of phage particles and their assembly, and mechanisms of bacterial cell immunity and its viral subversion. Modern techniques in genomics, proteomics, and structural biology are adding novel insights even now [13, 15]. This combination of precision studies illuminates the plasticity of macromolecular and cellular biology in this perennial cauldron of evolutionary cat and mouse.

In terms of replicative strategies, phage activity in a cell involves an interesting sideway taken on biological processes. The enzymatic agents encoded by phages have applications in laboratory research technologies, applied molecular evolution techniques, and also novel ways to make DNA for therapeutic and technological use: free of the heretofore necessity for bacterial cell passage to propagate this DNA and thus of bacterial propagation sequences, including antibiotic resistance genes. The insights from phages therefore bring us to the brink of revolutionising applications as diverse as human gene therapy, cell and regenerative medicine, and DNA as data storage (**Figure 1**).

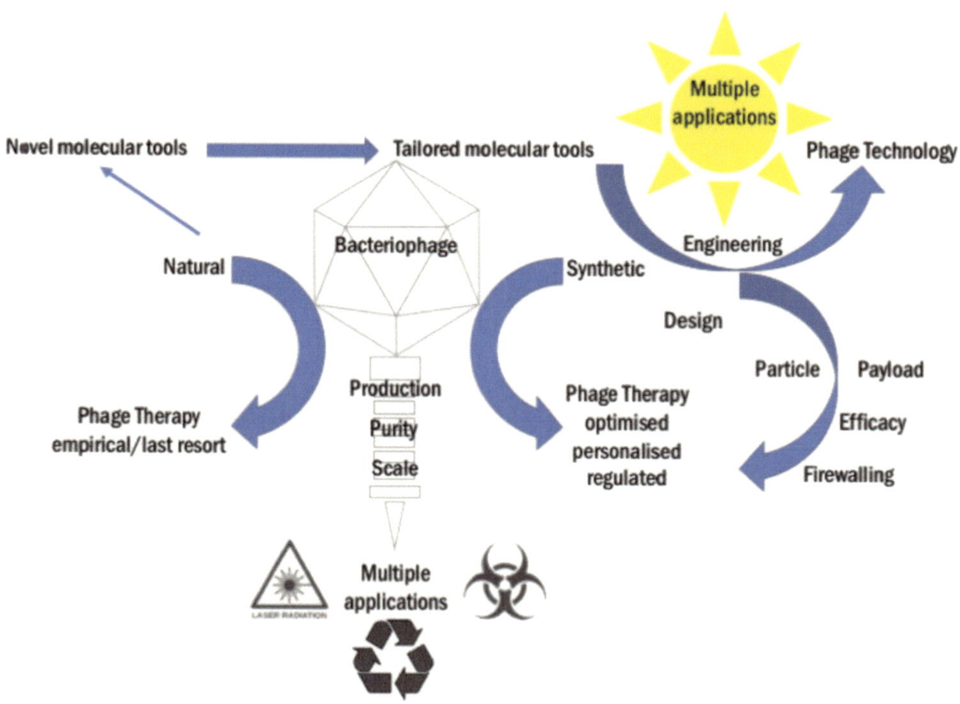

Figure 1.
Bacteriophages can be envisaged as natural microbial control agents, as well as machines for targeted synthetic genetic programming. The encoded proteins, as well as the structures of phages, offer a multitude of possibilities as outlined in this introductory chapter and detailed in the volume.

4. Rebooted

Finally, looking beyond what we have discovered and tested, beyond this volume and into that potential future of synthetically modified bacteriophages with diverse uses, what might we find? Perhaps we might find nanomachines inspired in their design by the multifarious forms and mechanisms of both known and newly studied types of these viruses, lightning-fast genotyping of bacterial infections at point of care, and efficiently timely synthetic tooling of medically approved phages to rapidly quell those infections, even on the scale of a few hours. We might also find synthetic phages reprogramming the microbiome, both by selective population control and by targeted and firewalled genetic modification in situ, and even the possibility of tooling wild species of bacteria for bioremediation purposes via reversible and firewalled genetic modification. There could be phage-encoded elements brought together in new and as yet unimagined combinations to effect all manner of building and alteration performed at the macromolecular scale. The future is always imagined yet unseen: with phages in mind that refers both to its dazzling scale of possibility and in its infinitesimal scale of operation. The bacteriophages will be just as much a major part of that synthetic future, as they have ever seemingly been in nature to this day.

Conflict of interest

Author Sophie E. Knott is a co-author of the included chapter *The unusual linear plasmid generating systems of prokaryotes* and has recently started a collaborative research partly funded by Touchlight Genetics Ltd., towards a PhD in the editor's laboratory: This fact presents no conflicts of interest to the presently published work.

Author details

Renos Savva
Institute of Structural and Molecular Biology, Department of Biological Sciences, Birkbeck, University of London, London, UK

*Address all correspondence to: r.savva@mail.cryst.bbk.ac.uk

References

[1] Clokie MRJ, Millard AD, Letarov AV, Heaphy S. Phages in nature. Bacteriophage. 2011;**1**(1):31-45. DOI: 10.4161/bact.1.1.14942

[2] Pendleton JN, Gorman SP, Gilmore BF. Clinical relevance of the ESKAPE pathogens. Expert Review of Anti-Infective Therapy. 2013;**11**(3): 297-308. DOI: 10.1586/eri.13.12

[3] Myelnikov D. An alternative cure: The adoption and survival of bacteriophage therapy in the USSR, 1922–1955. Journal of the History of Medicine and Allied Sciences. 2018; **73**(4):385-411. DOI: 10.1093/jhmas/jry024

[4] Morozova VV, Vlassov VV, Tikunova NV. Applications of bacteriophages in the treatment of localized infections in humans. Frontiers in Microbiology. 2018;**9**:1696. DOI: 10.3389/fmicb.2018.01696

[5] Parfitt T. Georgia: An unlikely stronghold for bacteriophage therapy. Lancet. 2005;**365**(9478):2166-2167. DOI: 10.1016/S0140-6736(05)66759-1

[6] Dedrick RM, Guerrero-Bustamante CA, Garlena RA, et al. Engineered bacteriophages for treatment of a patient with a disseminated drug-resistant *Mycobacterium abscessus*. Nature Medicine. 2019;**25**:730-733. DOI: 10.1038/s41591-019-0437-z

[7] Schmidt C. Phage therapy's latest makeover. Nature Biotechnology. 2019; **37**:581-586. DOI: 10.1038/s41587-019-0133-z

[8] Malika DJ, Sokolova IJ, Vinner GK, et al. Formulation, stabilisation and encapsulation of bacteriophage for phage therapy. Advances in Colloid and Interface Science. 2017;**249**: 100-133. DOI: 10.1016/j.cis.2017.05.014

[9] Krysiak-Baltyn K, Martin GJO, Gras SL. Computational modelling of large scale phage production using a two-stage batch process. Pharmaceuticals (Basel). 2018;**11**(2):31. DOI: 10.3390/ph11020031

[10] Kilcher S, Studer P, Muessner C, Klumpp J, Loessner MJ. Cross-genus rebooting of custom-made, synthetic bacteriophage genomes in L-form bacteria. Proceedings of the National Academy of Sciences of the United States of America. 2018;**115**(3):567-572. DOI: 10.1073/pnas.1714658115

[11] Ando H, Lemire S, Pires DP, Lu TK. Engineering modular viral scaffolds for targeted bacterial population editing. Cell Systems. 2015;**1**(3):187-196. DOI: 10.1016/j.cels.2015.08.013

[12] Yosef I, Goren MG, Globus R, Molshanski-Mor S, Qimron U. Extending the host range of bacteriophage particles for DNA transduction. Molecular Cell. 2017; **66**(5):721-728. DOI: 10.1016/j.molcel.2017.04.025

[13] Xu J, Wang D, Gui M, Xiang Y. Structural assembly of the tailed bacteriophage φ29. Nature Communications. 2019;**10**:2366. DOI: 10.1038/s41467-019-10272-3

[14] Lima-Mendez G, Toussaint A, Leplae R. Analysis of the phage sequence space: The benefit of structured information. Virology. 2007; **365**:241-249. DOI: 10.1016/j.virol.2007.03.047

[15] Hrebík D, Štveráková D, Škubník K, Füzik T, Pantůček R, Plevka P. Structure and genome ejection mechanism of Staphylococcus aureus phage P68. Science Advances.2019;**5**:eaaw7414. DOI: 10.1126/sciadv.aaw7414

Chapter 2

Bacteriophages: Their Structural Organisation and Function

Helen E. White and Elena V. Orlova

Abstract

Viruses are infectious particles that exist in a huge variety of forms and infect practically all living systems: animals, plants, insects and bacteria. Viruses that infect and use bacterial resources are classified as bacteriophages (or phages) and represent the most abundant life form on Earth. A phage can be described as a specific type of nano-machine that is able to recognise its environment, find a host cell, start infection, self-assemble and safeguard its genome until the next cycle of replication is initiated. Remarkable results have been obtained by combining cryo-EM, X-ray analysis and bioinformatics in structural studies of these nano-machines. In this review we will describe results of structural studies of phages that uncover their organisation in different conformations, thus facilitating our understanding of the functional mechanisms in supramolecular assemblies and helping us understand the usage of phages in medical treatments. Currently, antibiotic resistance is an enormous challenge that we face. The tailed phages could be used in place of antibiotics due to their high specificity to host cells, but more knowledge of their organisation and function is required.

Keywords: viruses, bacteriophage, structural organisation, infectivity, function, structural methods, electron microscopy

1. Introduction

All living systems have many diseases that are often caused by small organisms such as bacteria or infectious particles consisting of proteins, nucleic acids and sometimes lipids. These particles are called viruses, use the resources of living cells for their own propagation and can be transmitted from one organism to another. Each type of particle infects its own host cells, and they can survive outside living organisms in very harsh conditions. some of them continue to replicate with cells despite the host's defence mechanisms and remain dormant (latent) in their host cell, e.g. herpesviruses which reactivate at a later date to produce further attacks of the disease if the host's defence system weakens [1].

Bacteriophages (or phages) are viruses that infect and use bacterial resources for their own reproduction. They are characterised by a high specificity to bacteria at infection and are very common in all environments. Their number is directly related to the number of bacteria present. It is estimated that there are more than 10^{30} tailed phages in the biosphere [2]. Phages are common in soil and readily isolated from faeces and sewage, as well as being very abundant in freshwater and oceans with an estimate of more than 10 million virus-like particles in 1 mL of seawater [3, 4].

Why study the structure-function relationship of phages? Currently, there are substantial problems with diseases caused by bacteria, especially in hospitals.

Many pathogenic bacteria exist such as *Mycobacterium tuberculosis*, *Enterococcus faecalis*, *Staphylococcus aureus*, *Acinetobacter baumannii*, *Pseudomonas aeruginosa* and methicillin-resistant *S. aureus* (MRSA) and have become modified in hospitals due to the overuse of antibiotics. Bacteria have become resistant to some of the most potent drugs used in modern medicine, and this causes treatment problems [5–7]. It appears that the pathogenic bacteria adapt quicker to antibiotics than the new ones that can be produced. The number of new antibiotics being introduced has decreased since their first introduction [8].

A powerful method to circumvent this resistance is the use of phages in the treatment of bacterial infections [9]. Most current studies of phage therapy have focussed on acute infections in animals [10]. In order to regulate the mechanisms of phage infection, we need to know not only the phage structure but also the phage-cell surface interaction mechanism and the process of switching the cell replication machinery for phage propagation. One important factor that has to be considered is how phages are reproduced. Phages have two ways of propagation: lytic and lysogenic [11]. In the first case, phages cause the compete lysis of a cell, where it breaks open and subsequently dies after phage replication. In the second type of replication, a phage integrates its genome into the host bacterium's genome or forms a circular replicon in the bacterial cytoplasm. The bacterium then continues to live and reproduce normally, but the phage genome is transmitted to progeny cells at each subsequent cell division. Changes in cell conditions such as radiation or certain chemicals can release the phage genome, causing proliferation of new phages via the lytic cycle. Therefore, for medical treatments we need to use only lytic phages, so they will exist in an organism, while the pathogenic bacteria are around but only infect those bacteria that have the appropriate receptors in the outer membrane. This is an important factor that can be used to affect specific bacteria without harming those ones that are essential for the health of humans and animals [10]. In this review we will focus on tailed phages as they are abundant and well studied and could be beneficial to medicine [12]. We will describe the general organisation and structural features of their components revealed by current structural methods.

2. Phages and their classification

Virus classification is based on characteristics such as morphology, type of nucleic acid, replication mode, host organism and type of disease. The International Committee on Taxonomy of Viruses (ICTV) has produced an ordered system for classifying viruses (https://talk.ictvonline.org/taxonomy/). Phages are found in a variety of morphologies: filamentous phages, phages with a lipid-containing envelope and phages with lipids in the particle shell (**Figure 1A**). They have a genome, either DNA or RNA, which can be single or double stranded, and contain information on the proteins that constitute the particles, additional proteins that are responsible for switching cell molecular metabolism in favour of viruses and, therefore, the information on the self-assembly process. The genome can be one or multipartite and is located inside the phage capsid. Nearly 5500 bacterial viruses have been characterised by electron microscopy (EM) [15]. The shape of viruses is closely related to their genome, and a large genome indicates a large capsid and therefore a more complex organisation. The most studied group of phages is the tailed phages (order *Caudovirales*) which are classified by the type of tail; *Siphoviridae* have a long non-contractile tail, *Podoviridae* have a short non-contractile tail and *Myoviridae* have a complex contractile tail (**Figure 1B**).

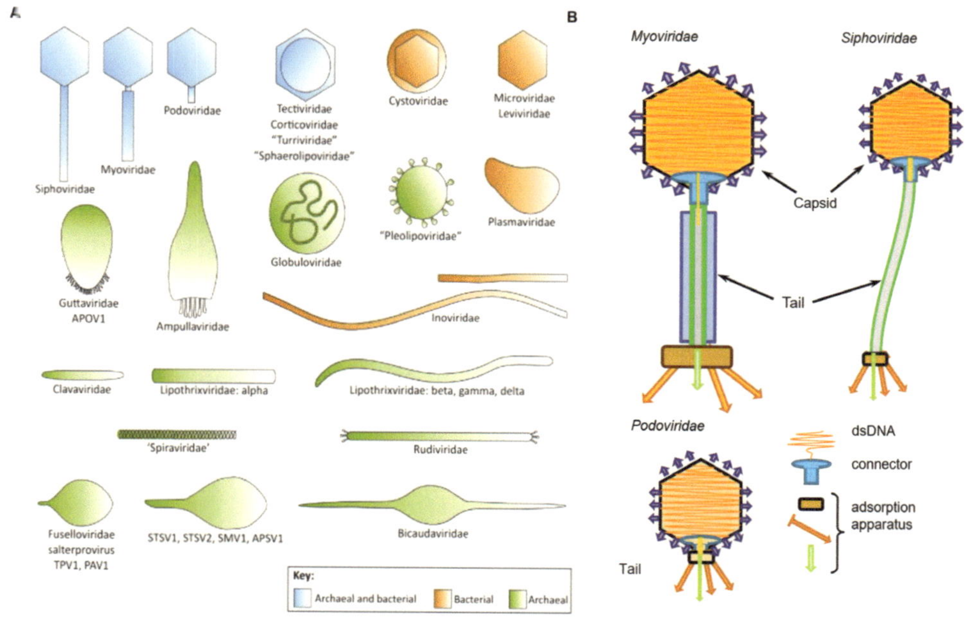

Figure 1.
(A) Representation of prokaryote bacteriophage morphotypes [13]. (B) Members of the Caudovirales family [14].

3. Methods used for structural studies of viruses

The first ideas on how viruses infect cells were based on results obtained by microbiology and bacteriology during the last century. Understanding the function of viruses and how this can be regulated and modified requires knowledge of their structural organisation. However, investigation of structure-function relationships needs a combination of different techniques. Microbiology has identified viruses as infectious agents, while bacteriology and light microscopy enabled us to identify specificity between viruses and host cell interactions and to recognise a level of survival of bacteria in the presence of different phages. In order to understand interactions at the molecular level, one needs to know the structural features of the viruses and their components at an atomic level. Different structural techniques are often utilised for smaller components, and the results fitted into larger EM structures.

3.1 X-ray crystallography

X-ray crystallography was the first method used to study proteins at the atomic level, which is essential to reveal protein-ligand interactions that can boost or suppress protein activity. It is based on the principles of beam scattering within a crystal. By using specific software packages, a 3D electron density map of the protein that forms the crystal can be calculated [16]. However, to produce protein crystals, we need solutions of a protein at high concentration. The proteins have to be stable, and often mutations are made to remove their flexible parts, but this may produce different conformations to those that are required for their natural activity.

X-ray analysis is an efficient tool for analysis of protein complexes from a few kDa to hundreds of kDa in size. In order to study the structure of a large protein or a complex of several proteins, the process of crystallisation becomes a more challenging step. The development of cryoprotection in X-ray crystallography, where the crystals are flash frozen, has improved the quality of the data and often resulted in higher resolution. Nowadays, many structures of large protein complexes (up to

2–3 MDa) have been determined by X-ray analysis, but these projects have required decades to obtain high-quality crystals [17].

Viruses are much bigger particles and often have flexible components. The large size of the complexes results in significantly bigger unit cells, which results in technical challenges in obtaining fine structural details. Viruses with a rigid icosahedral lattice of the capsid have been studied successfully by X-ray crystallography at near-atomic resolution. The first viral structure was that of the *Blue tongue virus* (700 Å diameter) determined at a resolution of 3.5 Å which was the largest virus structure determined at that time [18]. The capsid of the *Siphoviridae* phage HK97 (without a portal protein) was determined at a resolution of 3.5 Å [19]. Later studies have shown that the fold of the HK97 phage capsid protein, which forms the envelope to protect viral genome from the harsh outer environment, represents a conservative fold found nowadays in nearly all dsDNA viruses so far studied.

3.2 Nuclear magnetic resonance

Nuclear magnetic resonance (NMR) is an important technique that resolves structures of small proteins that are not suitable for crystallisation due to their flexibility. This method is based on exploiting the electrical charges and spins of the nuclei in a molecule. If an external magnetic field is applied, energy is transferred to the nuclei changing their state from the level of base energy to a higher energy. This energy is emitted when the spin returns back to its base level at a frequency corresponding to radio frequencies[1]. The signal that matches this transfer is measured and processed in order to yield a NMR spectrum [17, 20]. This technique is typically used for proteins of less than 200 amino acids and an upper weight limit of about 50 kDa, so it is unsuitable for the structural determination of complete viruses. However, it can be used to analyse flexibility of bigger complexes [21]. The NMR structures can be docked into low-resolution cryo-EM structures.

3.3 Electron microscopy

Light microscopy has been used for several centuries to study objects that are hardly visible to the naked eye. In conventional microscopy, resolution is mostly restricted according to the theoretical context of the Rayleigh criterion [22]. This limit is defined by the diffraction properties of light in lenses and has restricted our view to objects bigger than 250 nm. New developments in technology and advances in optical quality, electronics and software have delivered new options and extended the field of applications for electron microscopes allowing visualisation of single molecules. Electron microscopes use a beam of electrons (wavelength of less than 0.1 nm) instead of visible light (wavelength 400–700 nm). Due to their charge, the electrons can be focused using an electromagnetic field, which is why the optical system of the electron microscope (EM) is similar to the general optical system of light [23]. The short wavelength of the electron beam allows details of small objects less than 0.1 nm in size to be seen. However, biological samples are not stable in the vacuum necessary to create an image using electrons that would otherwise become absorbed by air, and, moreover, biological samples are sensitive to the electron irradiation. These factors reduce the level of achievable resolution.

At the very beginning of EM evolvement, a method called negative staining was used for visualisation of biological complexes. In this case a drop of biocomplex solution is placed on a support grid and embedded in a heavy atom salt, usually uranyl acetate [24]. Since the specific density of the negative stain is much higher than

[1] chem.ch.huji.ac.il/nmr/whatisnmr/whatisnmr.html

the density of the biological molecules in the microscope, we can see the cast of the molecule merged into the surrounding stain. Where the stain did not penetrate into the molecule, one can see light spots in the image as the stain has blocked electrons. Sample preparation is fast and produces very high contrast. However, this technique does not allow fine details to be seen, and the particle becomes distorted due to the drying procedure required. The stain has a relatively large grain (up to 1.5 nm) that obscures details of the molecules under study.

Nearly four decades ago, a cryo-technique for sample preparation was introduced that allows biocomplexes to be kept at nearly native conditions. A thin layer of sample on a grid is flash frozen at liquid nitrogen temperatures, thus trapping molecules in a native, hydrated state within a thin layer of amorphous ice [25]. This technique is used to study the structural organisation of biocomplexes by cryo-electron microscopy (cryo-EM) or electron tomography (cryo-ET). Until two decades ago, all data in EM was collected on films that had to be developed and digitised, which was time-consuming. The advent of charge-coupled devices (CCDs) allowed direct digital acquisition of images and the collection of large numbers of particles giving rise to structures of higher resolution. Later, direct electron detectors were introduced into EM and are now used in all high-end electron microscopes [26]. Together with new approaches in microtechnology and the automation of data collection, the results from image analysis have improved tremendously. Cryo-EM is now approaching the near-atomic resolution that had only been achieved by X-ray crystallography. New maps obtained by cryo-EM provide information on the main polypeptide chains and often reveal the positions of side chains. The current highest resolution of structures currently deposited in the EMDB is 1.5 Å [27], with many others at a resolution between 3.5 and 4 Å. At this resolution atomic models can be built and refined using the crystallographic methods.

In cryo-ET the samples are also flash frozen, but data is collected by tilting the grid with the sample between −60 and 60° around the horizontal axis (perpendicular to the optical axis of the microscope) with an increment typically of 2°. The 2D images taken at each angle are combined to calculate a 3D map of the object. The limitation in the range of the tilt results in a cone of missing data [28]. The resolution in structures obtained by cryo-ET is lower than that in single-particle analysis. However, this approach allows visualisation of important organelles within cells. If there are multiple small structures such as ribosomes or viruses, then each structure can be extracted and averaged. This is called subtomogram averaging and will give higher-resolution structures [29].

4. Overall structural organisation of phages

Phages may have different shapes and sizes (**Figure 1A**). The most studied group is that of tailed phages with a dsDNA genome, and it also represents the largest group (**Figure 1B**). The tailed phages have three major components: a capsid where the genome is packed, a tail that serves as a pipe during infection to secure transfer of genome into host cell and a special adhesive system (adsorption apparatus) at the very end of the tail that will recognise the host cell and penetrate its wall. Cell resources are used for the phage reproduction.

The functional phage is a result of a multistep process that starts with all the necessary proteins produced by the host cell after infection: capsid, portal, tail, scaffolding, terminase, etc. (**Figure 2**). The capsids of the dsDNA phages often have fivefold or icosahedral symmetries [30], which are broken at one of the fivefold axes by the head-to-tail interface (HTI). The main component of the HTI is a dodecameric portal protein (PP) within the capsid. The PP represents the DNA-packaging motor,

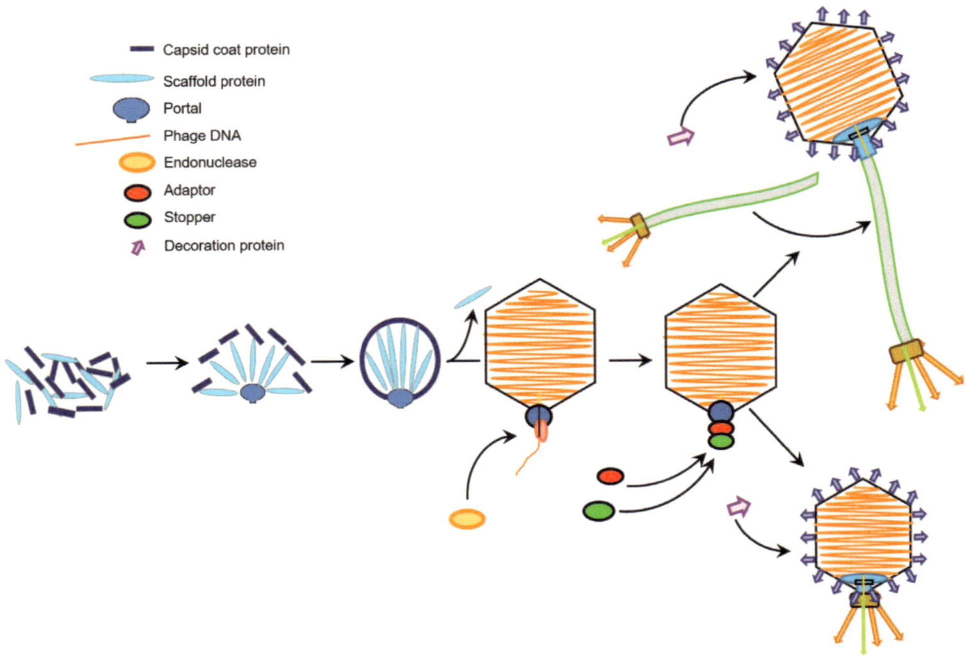

Figure 2.
Self-assembly pathway of phages. Multiple copies of the capsid/scaffold complex bind the portal protein to form the procapsid; then, the scaffold proteins are ejected, and DNA is packaged into the procapsid, which expands to the size of the mature capsid. The head completion proteins (the stopper and the adaptor) are bound to the portal complex preventing DNA leakage. Next, decoration proteins bind to the capsid, and the tail, assembled separately or after DNA packaging, is attached; thus, the final infectious phage is produced. The preassembled tail attaches in Myoviridae and Siphoviridae, while in Podoviridae the tail assembles at the stopper.

which is the crucial part of these nano-machines. The HTI also includes oligomeric rings of head completion proteins that play dual roles: (1) making an additional interface to molecules of ATP which provide energy for DNA packaging and (2) then connecting the portal protein and the tail. Some HTIs also serve as valves that close the exit channel preventing leakage of genome from the capsid but opening as soon as the phage is attached to the host cell. However, symmetries other than dodecameric have been found for nearly all PPs in vitro if the PPs are assembled under naive conditions, without any other phage protein components [31–35]. Typically, the main phage proteins have conservative folds despite low sequence similarity, although they may have different additional domains [36, 37].

The phage tail is the structural component of the phage that is essential during infection. Its adsorption apparatus located on the distal end of the tail recognises a receptor, or the envelope chemistry, of the host cell and ensures genome delivery to the cell cytoplasm. In *Myoviridae* and *Siphoviridae*, the tail is composed of a series of stacked rings with the host recognition device being located at the end of the tail. In *Podoviridae* the adsorption apparatus is bound immediately to the HTI. The adsorption apparatus is surrounded in many phages by fibrils that ensure a tight connection to the host cell (**Figure 2**).

4.1 Procapsids

The capsid of a phage has a precursor formation, named the procapsid, during the assembly process (**Figure 2**). Scaffolding proteins (SPs) drive the assembly process by chaperoning major capsid protein (MCP) subunits to build an icosahedral procapsid that is later filled with dsDNA. The SPs are bound to the portal

Phage	Type of phage	Capsid protein	No. of residues	M. Mass (kDa)	Resolution (Å)	Structure analysis
HK97	*Sipho*	gp5	385 282 (AC)	42	3.44 (C) 12 (PC)	X-ray [42] EM [51]
T5	*Sipho*	pb8	458 299 (AC)	51	9 (C)	EM [52]
λ	*Sipho*	gpE	341	38	6.8 (C),13.3 (PC)	EM [47]
SPP1	*Sipho*	gp13	324	35	8.8 (C)	EM [53]
TP901-1	*Sipho*	ORF36	272	29	15	EM [54]
TW1	*Sipho*	gp57*	352	39	7	EM [55]
φ29	*Podo*	gp8	448	50	8	EM [56]
T7	*Podo*	gp10	345	37	4.6 (PC) 3.6 (C)	EM [57]
P22	*Podo*	gp5	430	47	3.8 (PC) 4.0 (C) 3.3 (C)	EM [58–60]
ε15	*Podo*	gp7	335	37	4.5	EM [50]
T4	*Myo*	gp24 gp24* gp23 gp23*	427 417 (AC) 521 456 (AC)	47 44 56 49	2.9 (monomer) 3.3 (EM)	X-ray [61] EM [62]
HSV-1	*virus*	VP5	1374	149	4.2 (C)	EM [63]

AC—after cleavage; C—capsid; PC—procapsid

Table 1.
Phage procapsids and capsids.

complex during formation of a procapsid with scaffolding inside. The sequence of conformational changes from a procapsid to the phage capsid where genome has been packed is named as the maturation process and goes through a series of intermediates [19, 38–40]. Some phages like HK97 and T5 do not have a separate SP; instead, the capsid protein is fused with a scaffolding domain at the N-terminus. As soon as the procapsid is assembled, the scaffolding domain is cleaved off and then like the separate SP will be removed from the capsid to make room for the genome [38, 39]. Structures of procapsids and mature virions have been determined for a number of phages (**Table 1**). The spherical capsid shell expands during maturation and becomes thinner due to alterations in the inter- and intra-subunit contacts.

4.2 Capsids

Most tailed phages have capsids of an icosahedral shape formed by multiple copies of one or more proteins. Icosahedral capsids are characterised by 12× fivefold, 20× threefold and 30× twofold axes, which give rise to 60 copies of the major independent parts [41]. A triangulation number (T number) describes the number of copies of the same protein within the independent part of the icosahedral lattice. The overall number of proteins in the virus corresponds to the T number multiplied by 60; for example, a T = 3 virus has 180 subunits [41]. Oligomers of the proteins that are located on the fivefold axes are referred to as pentons, while those complexes that are located on the faces of the icosahedron and form oligomers from six subunits are named as hexons.

***HK97** phage.* The first structure of a phage capsid was solved for the *Siphoviridae* **HK97** phage (a mutant without the PP and tail, diameter ~650 Å, T = 7) at 3.6 Å resolution [19] by X-ray crystallography. This structure revealed a new type of protein fold, which has been found in many other phage capsid proteins (CP) despite low sequence identity (**Table 1**). Later, the structure was improved to 3.45 Å resolution [42]. This fold is also found in distantly related icosahedral tailed viruses that infect halophilic archaea [43] and human pathogens such as *Herpesviridae* [44–46]. The characteristic features of the HK97-fold are an N-arm, a peripheral P-domain with a long helix and a β-sheet, an axial A-domain and a long E-loop that fills the region between adjacent threefold axes (**Figure 3A**). The HK97 capsid is held together by molecular chain mail [19]. The two antiparallel β-strands in the E-loop are terminated by a loop, containing Lys169. This residue forms an isopeptide bond with the P-domain residue Asn356 of an adjacent subunit. The third residue involved is the catalytic residue Glu363 from a third subunit (**Figure 3A**). Conformational changes between HK97 procapsids and capsids were assessed by flexible fitting of atomic models [40]. The capsid subunits are skewed around a pseudo-twofold axis in the procapsid but form more symmetric pentons and hexons in the mature virion.

λ phage. The mature capsid of the *Siphoviridae* phage λ (diameter ~600 Å, T = 7) is stabilised with the help of a decoration protein gpD which is attached to the threefold vertices. A procapsid structure was determined with a resolution between 13.3 and 14.5 Å and a mature capsid at 6.8 Å [47]. The capsid protein gpE of the mature capsid shows clearly the HK97-fold so the HK97 model was used for rigid body fitting into the capsid [47]. However, the A-domain gave a poor fit so it was treated as a separate rigid body and was found to be rotated 15° clockwise compared with HK97. There was unassigned density, which could fit the extra 59 residues in

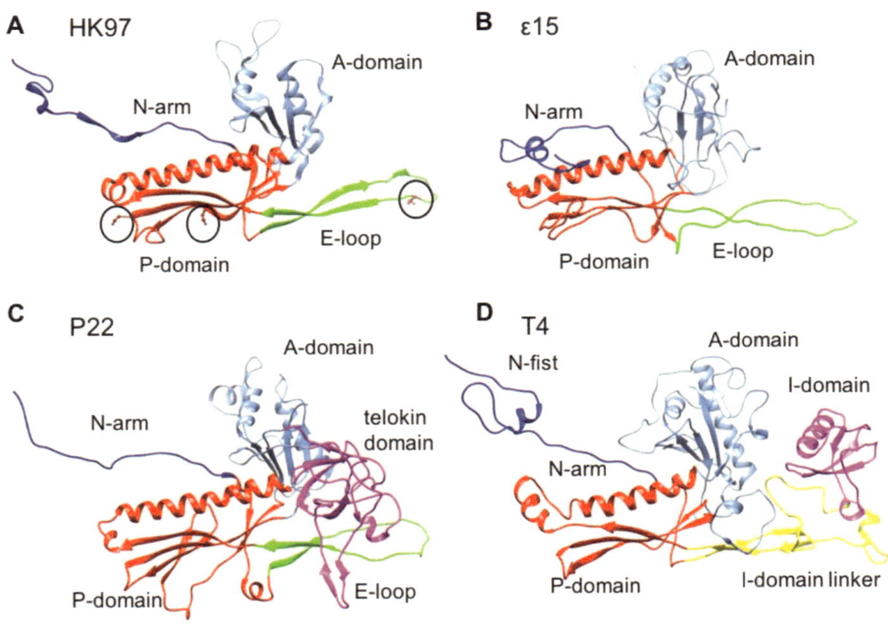

Figure 3.
Structural organisation of the major capsid proteins. (A) Siphophage HK97 (1OHG). The catalytic residues are shown in brown and circled. (B) Podophage ε15 (3J40). (C) Podophage P22 (5UU5). (D) MyophageT4 gp23 (5VF3). The N-arm is dark blue, the P-domain is red, the A-domain is light blue and the E-loop is yellow. Extra inserted domains seen in P22 and T4 are magenta. The yellow linker in T4 is topologically equivalent to the E-loop seen in the other phages.*

the λ phage. The crystal structure of gpD (1.1 Å) [48] was fitted into the cryo-EM capsid map and, when combined with the HK97 model, showed how gpD is held in place at the capsid threefold axis.

ε15 phage. The structure of gp7, the MCP of the *Salmonella* phage **ε15**, was determined by cryo-EM (4.5 Å) [49] and showed a fold similar to that in HK97 (**Figure 3B**). Later, the same group obtained a better resolution structure (~3.5 Å) [50] and, using their previous 4.5 Å structure for gp7, computationally analysed possible different gp7 conformations within EM density obtained from averaging the individual subunits in one asymmetric unit. They found that two models could be fitted in the structure: one that was consistent with the previous interpretation [49], and the other had a strand swap of the β-strands in the P-domain before the two helices in the A-domain. The strand order was 4-3-1-5-2, but in the swapped model, it was now 5-4-2-3-1. The model with the strand swap provided a better visual fit in the refined map [50].

P22 phage. Structures of a procapsid and a heat-expanded capsid of the *Podoviridae* **P22** phage (diameter ~620 Å, T = 7) were determined by cryo-EM at 9.1 and 8.2 Å, respectively [58]. This heat-expanded capsid was composed only of hexons with wide openings where the pentons would be. The MCP gp5 revealed the HK97-fold, but there was an extra domain (residues 223–349) with an immunoglobulin-like telokin domain [64]. Later, the procapsid and capsid structures were determined by cryo-EM at 3.8 and 4.0 Å resolution, respectively [59]. It was also found that gp5 has the HK97-fold with extra density above the E-loop corresponding to the telokin domain (**Figure 3C**). A model built from a cryo-EM map of P22 (3.3 Å) [60] identified the interactions that stabilise the capsid. In the mature virion, the N-arm forms an antiparallel β-strand pair between neighbouring subunits around the threefold axis. A second set of interactions involve hydrogen bonds and salt bridges between adjacent subunits involving the A-domain, the E-loop and the I-domain [60].

φ29 phage. The podovirus **φ29** is a relatively small phage (prolate icosahedra, height ~540 Å, diameter ~450 Å) that requires pRNA along with the ATPase gp16 to provide enough energy for the DNA translocation. The procapsid consists of the MCP gp8, the SP gp7, the head fibre protein gp8.5, the connector gp10 and a pRNA. After DNA packaging, the pRNA and ATPase come off the procapsid and are replaced by the gp11 and gp13 underneath the connector, gp29 at the end of the tail and the appendages (gp12*) which are attached to gp11 and gp13 [65]. During maturation an 18 kDa fragment of gp12 is cleaved off to give gp12* found in the appendages whose role is to adsorb the virion on the host cell. A fibreless, isometric φ29 variant capsid cryo-EM structure (7.9 Å) showed that gp8 has a structure with the HK97-fold but with extra density [66]. Domain profile-searching algorithms [67] showed that residues 348–429 at the C-terminus were found to be 32% identical to a BIG2 domain consensus sequence (group 2 bacterial immunoglobulin-like domains). An asymmetric reconstruction of fibreless full (7.8 Å) and empty (9.3 Å) capsids by cryo-EM revealed the interactions between the capsid shell and DNA [56]. All these interactions were in similar or identical locations in most of the gp8 subunits. The most prominent contact was at the end of a long tubular piece of density, which after fitting a HK97 homology model could be assigned to the N-terminal end of the HK97 long helix.

T4 phage. The protein elements of the *Myoviridae* phage **T4** (prolate icosahedra, height ~2000 Å, diameter ~900 Å) and its overall organisation have been extensively studied by X-ray and EM [62, 68–70]. The procapsid of T4 contains two proteins gp23 and gp24 which have 22% sequence identity. During maturation the gp21 protease cleaves off 65 N-terminal residues from the capsid protein gp23 and 10 N-terminal residues from the vertex protein gp24 to produce gp23* and gp24*, respectively. In the mature capsid, gp23* forms 120 hexons, and 11 capsid vertices are formed by gp24* proteins, while the 12th is occupied by a dodecamer of gp20 PP. T4 has two decoration proteins Soc and Hoc [62]. A crystal structure of gp24 (2.9 Å) [61] showed a domain with the HK97-fold and a 60 residue insertion I-domain located on the outer capsid surface (**Figure 3D**). The 3.3 Å cryo-EM reconstruction of the isometric capsid of T4 [62] allowed the structure of gp23* and gp24* to be determined. The I-domain linker, missing from the crystal structure of gp24, could be seen and interacts with a neighbouring gp24* molecule to stabilise the capsid. The structure of gp23* is similar to gp24* but with an extra compact region formed by residues 66–93, termed the "N-fist" prior to the N-arm residues 94–110 [62].

Crystal structures were obtained for the Hoc protein from the T4-like phage RB49 with the capsid-binding C-terminal domain 4 missing [71] and Soc protein from the T4-like phage RB49 [72]. The Soc molecules, which are required for capsid stability, interact with three gp23* subunits [62] although not all binding sites were fully occupied possibly due to differences in the gp23* I-domain linkers. The immunogenic outer capsid Hoc protein was found in two different sites within the asymmetric unit: at the centre of the hexon near the icosahedral threefold axis and in the hexon close to the fivefold axis [62]. The density of Hoc near the threefold axis was less interpretable than that near the fivefold axis.

***HSV-1** virus*. Although it is not a phage, the human herpesvirus **HSV-1** capsid (1250 Å, T = 16) is a close relative and undergoes a pathway of self-assembly similar to that of dsDNA phages [73]. The virion is characterised by the following features: envelope, tegument, capsid and the viral genome. There are three types of HSV capsids: A-capsids have neither DNA nor the SP, B-capsids have the SP but no DNA and C-capsids contain DNA but no SP. The MCP VP5 forms pentons and hexons and VP26 binds to the VP5 hexons. A triplex of VP19C and VP23 found between capsomers [45]. The upper domain of residues 451–1054 was crystallised and the structure was determined at 2.9 Å. The structure of the whole virion of HSV-1 was determined at ~7 Å resolution [45]. The model of HK97 capsid protein was fitted to the lower domain of VP5 where the E-loop and N-arm were visible, the spline helix was longer and the central channel was wider. Unlike HK97, the E-loop does not form the covalent cross-links or reach an adjacent capsomer. Instead, it interacts with adjacent subunits, lower and middle domains of same VP5 subunit and a triplex molecule. A structure of HSV-1 with its capsid-associated tegument complex (CATC) has been obtained at 4.2 Å [63]. VP5 had the HK97-fold with six additional domains. The two β-barrels in HSV Tri1 (VP19c) and Tri2 (VP23) resemble the homotrimers found in proteins like gpD of phage λ.

4.3 Connectors

In phages and herpesviruses, one of the fivefold vertices of the capsid is replaced by a *head-to-tail interface (*HTI*)* [30], which is a multi-protein complex (connector). In all phages the HTI provides a platform for docking of preassembled tails in *Sipho-* or *Myoviridae* or initiates the assembly of a short tail in *Podoviridae* [30]. The HTI comprises a portal complex (PP) and head completion proteins (**Figure 2**) that serve as a valve for closing the channel and keeping the phage genome inside the capsid at high pressure and only opens to allow genome release from the capsid (under natural conditions) as soon as the phage becomes tightly attached to a host cell.

All currently known PPs are homo-dodecamers when extracted from the viral capsids, as that symmetry is imposed during self-assembly in vivo. However, naive assemblies in vitro of the PP complexes have some variations in their rotational symmetry with 13-mers being observed for SPP1, T7 and HK97 [31, 33, 74]. HSV has been shown to have 11-fold, 12-fold, 13-fold and sometimes even 14-fold symmetry [34].

While monomers of the different PPs vary in size, all of them share a common fold—shown by EM and X-ray structures that were obtained for the φ29, SPP1 and P22 portals [75–78] and by cryo-EM for T7 and T4 (**Table 2**) [69, 79]. All known PP monomers are characterised by four domains: clip, stem, wing and crown (**Figure 4**) [77]. The clip domain is exposed to the capsid exterior and involved in binding to the terminase for DNA packaging [75, 80, 81] and later to a head completion protein during the HTI assembly [82]. The first high-resolution structure of a phage PP was obtained for the φ29 phage (**Figure 4A,** [75]). The clip domain is linked to the wing region through a stem that comprises typically two α-helices and the outer loops (**Figure 4B, C**). X-ray structures of PP from φ29 and SPP1 phages revealed major helical components that form the central channel through which DNA enters and exits the capsid. The structures of other PPs obtained later have confirmed that this is a conserved element characteristic for all known PPs. The wing domain radiates outwards from the central axis and has an α-helix, which is the longest one and serves as a spine of the wing. It has an α/β sub-fold at its periphery [77]. The crown domain consists of α-helices and is relatively small in SPP1 and surprisingly long (213 aa) in phage P22 (**Figure 4B, D, Table 2**).

***SPP1** phage*. The **SPP1 phage** connector has been studied for nearly two decades; the SPP1 PP structure was determined by X-ray crystallography, and all other portal complexes are compared with it (**Figures 4B, C** and **5A, B**), but this structure was a 13-mer [31]. The HTI of SPP1 extracted from the capsid was a stable complex composed of the PP gp6, the adaptor protein (AP) gp15 and the stopper (SP) gp16, all organised as three stacked cyclical homo-oligomers [82–84] (**Figure 5A, B**). The cryo-EM structure of the SPP1 HTI before and after DNA release was obtained by cryo-EM at ~7 Å resolution, where the HTI is bound to the tail [82] with gp16 acting as a docking platform for the SPP1 preassembled tail [85, 86]. Binding of the tail induces changes in the position of the gp16 residues Ile9 to Thr33 that close the central channel of the connector.

***HK97** phage.* There is no structure of the PP gp3 of the **HK97 phage**, but structures of gp6 and gp7 that correspond to the AP gp15 and SP gp16 of SPP1, respectively, were determined by X-ray analysis [74]. The 2.1 Å crystal structure of the gp6 AP revealed that it forms a 13-mer during crystallisation. A model for a

Phage	PP	No. of residues	M. Mass (kDa)	Resolution (Å)	Structure analysis
HK97	gp3	424	47		none
T5	pb7	403	45		none
λ	gpB	533	59		none
SPP1	gp6	503	57	3.4 (X-ray), ~7 (EM)	X-ray [77], EM [82]
TP901-1	ORF32	452	52	20	EM [54]
TW1	gp24	459	51	21	EM [55]
φ29	gp10	309	36	2.1	X-ray [76]
T7	gp8	536	59	8, 12	EM [87]
P22	gp1	725	83	10.5 (EM) 3.25 (X-ray)	EM [88] X-ray [78]
ε15	gp4	556	61	20	EM [89]
T4	gp20	524	61	3.6	EM [69]
HSV-1	pUL6	676	74.2	8	EM [90]

Table 2.
Phage portal proteins.

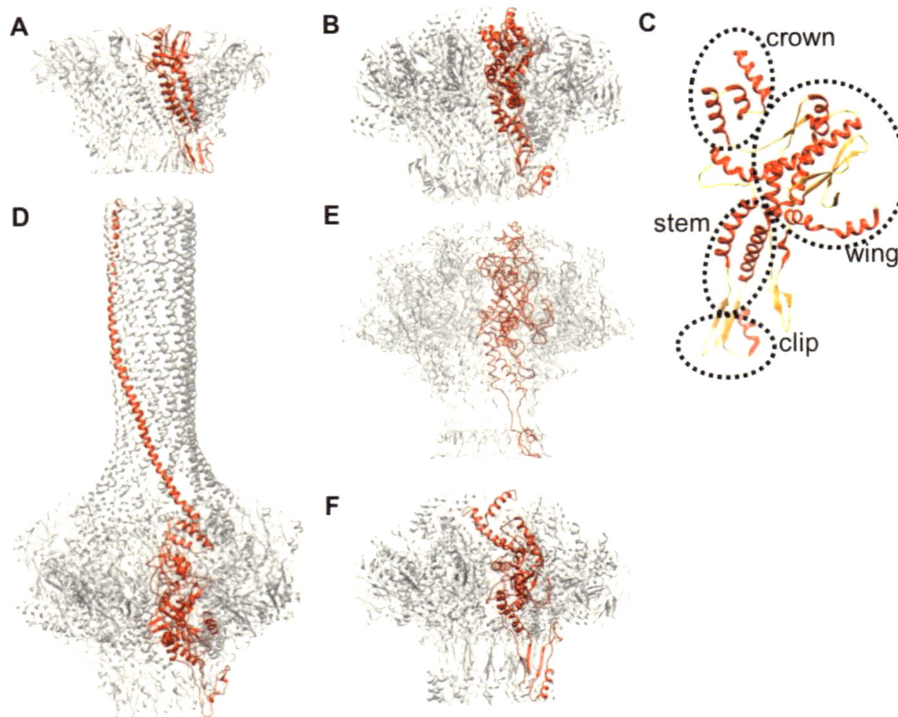

Figure 4.
Structures of portal proteins. One chain of the PP is highlighted in red. (A) gp10 of φ29 (1FOU). (B) gp6 of SPP1 (2JES). (C) A gp6 SPP1 monomer with the crown, stem, wing and clip domains indicated. (D) gp1 of P22 (3LJ5). (E) gp8 of T7 (3J4A). (F) gp24 of T4 (3JA7).

dodecameric ring of gp6 was constructed from a monomer taken from the 13-mer and fitted into a cryo-EM map of SPP1 [83]. This fitted well in size and shape, but the helices of HK97 gp6 did not fit well into the densities of the SPP1 connector EM map which suggested that the assembly into a 13-mer in the absence of other phage components may produce a different conformation [74].

***P22** phage*. The HTI of the **P22 phage** consists of two proteins, the PP gp1 and the AP gp4. The first structural organisation of the P22 HTI complex was obtained by cryo-EM at a resolution of 9.4 Å [91]. A crystal structure (3.25 Å) was obtained for the PP by using low-resolution EM data for phasing (**Figure 5C, D** [78]). The complete polypeptide chain was traced apart from a loop between residues 464 and 492, and loop modelling was used to build this area from a 9.2 Å cryo-EM map [88]. The overall height is ~300 Å with the gp4 AP forming a dodecameric ring below the PP. The PP of P22 has the same fold in its central channel as SPP1 [90] and φ29 [88, 89] (**Figures 4D, 5D**). The crystal structure of the full-length PP revealed that the C-terminal domain forms a ~200 Å long, α-helical barrel. At the same time, an asymmetric reconstruction of the entire P22 virion has been determined by cryo-EM at 7.8 Å resolution [92]. The 150 Å coiled-coil barrel structure extends from the PP to near the centre of the capsid. Fitting the crystal structure of the core-gp4 complex into the 7.8 Å virion density map revealed a stretch of about 21 gp4 C-terminal residues that lie wedged between the capsid and portal [92]. Overlap between gp4 and the MCP gp5 indicates that gp4 must undergo significant conformational change during phage assembly when the tail is added. A comparison of the portal position in the procapsid and the virion shows that the portal increases its contact with the capsid shell during maturation. It was proposed that a portion of the scaffold remains in place during dsDNA packaging to allow access of the gp4

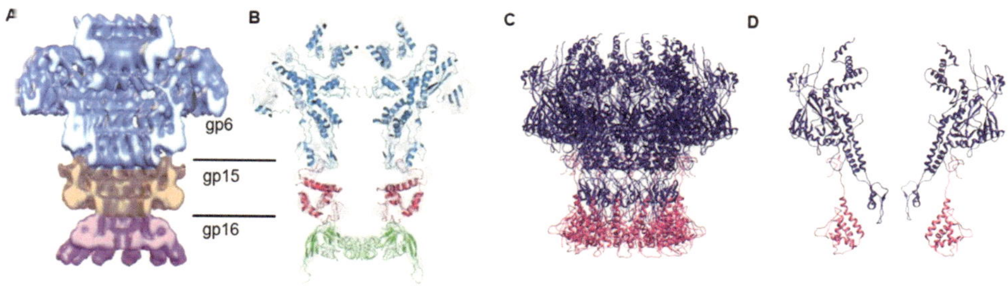

Figure 5.
Structures of the HTI. (A) The cryo-EM map of the SPP1 HTI coloured according to protein with the gp6 PP (blue), adaptor gp15 (brown) and stopper (purple) (EMD-1021 [83]). (B) Cutaway view of the SPP1 HTI with models gp6 (2JES), gp15 (2KBZ) and gp16 (2KCA) fitted into EMD-1021 (from [84]). (C) P22 connector complex determined by X-ray crystallography without barrel domain, PP pb1 (blue) and AP gp4 (red). (D) Central slice of C [91].

C-terminus to the bottom of the portal [92]. When gp4 binds, the SP is displaced allowing the final conformational change implied by the position gp4 C-terminal polypeptide that is wedged between the capsid and the HTI.

T7 phage. The HTI of **T7** was determined at 8 Å resolution [79], and the dodecameric T7 PP was found to be structurally similar to the PPs from other phages (**Figure 4E**). A ~12 Å resolution structure of a recombinant HTI (gp8-gp11-gp12) complex of T7 was determined by cryo-EM [87], and pseudo-atomic models were obtained using gp1 and gp4 of P22 for fitting and analysis of the T7 gp8 (PP) and gp11 (AP), respectively [78]. The T7 gp8 model was superimposed on gp10 from φ29 [75, 76], gp1 from P22 [78] and gp6 from SPP1 [77] which confirmed the presence of two stem helices, but the fold of the clip domain varies between the different PPs [87]. Previous structures of the head completion proteins have shown them to contain four helices, and when the T7 gp11 model was superimposed on gp4 from P22 and gp6 from HK97 [74, 78], the position of these four helices was conserved, but the C-termini were flexible.

T4 phage. The structure of the **T4** HTI region was determined by cryo-EM for the fully assembled capsid (~17 Å) [93]. The neck region, which connects the tail to the dodecameric PP (gp20), comprises adaptor proteins gp3, gp15, gp13, gp14 and gp *wac* (fibritin). It was assumed that the T4 HTI would have a similar PP organisation to other phages; therefore, the crystal structure of the φ29 PP was tentatively docked into the cryo-EM map of the T4 HTI region [75, 93]. Recently, the structure of the gp20 PP from T4 was determined by cryo-EM (3.6 Å) (**Figure 4F**, [69]). Interestingly, a gp20-N74 construct with the 73 N-terminal residues truncated contained 95% dodecamers and 5% 13-mers in solution [69]. The dodecameric PP was ~120 Å in height and varied in diameter from 90 Å (top) to 170 Å (middle) and to 80 Å close to the capsid, while the central channel goes from 90 to 28 Å near the middle of the channel. The connection to the capsid was by the PP wing domains, and the interactions are partially hydrophobic and partially polar [69].

φ29 phage. The asymmetric 3D reconstructions of ***φ29*** with and without DNA obtained by cryo-EM at 7.8 and 9.3 Å, respectively, showed that the PP gp10 dodecamer has elongated densities lining the central channel and tilted at ~30° to the central axis of the virion [56]. These cylindrical densities correspond to the α-helices seen in the crystal structure of the φ29 PP [76]. The density of the clip domain in the cryo-EM map lies further away from the PP axis than in the crystal structure possibly due to a different conformation of the PP within the fully assembled phage. Two cylindrical columns of high density were observed within the virions: one in the upper part and another at the bottom of the HTI. These densities were assigned to DNA based on the diameter, intensity and their location. The φ29 DNA is visible as ringlike densities below the PP, and then DNA stretches into the tail to about

~100 Å [56]. The DNA appears to contact with the PP crown domain and then with the density corresponding to the tunnel loops located between the PP crown and stem domains. These tunnel loops in the narrowest part of the connector channel were found in the SPP1 phage PP and are believed that they play a role in DNA translocation [77, 94]. The φ29 DNA appears to be clamped at the top of the tail tube [56].

***HSV-1** virus*. Herpes simplex virus (**HSV-1**) has its DNA packaged into the capsid through a portal channel of the PP complex (pUL6) [95]. The structure of a dodecameric HSV-1 PP has been determined by cryo-EM at 16 Å resolution from purified portals [34]. The structure showed a close resemblance to the SPP1 PP [83]. The PP is about the same size as the pentons that occupy the other fivefold vertices, which explains the difficulty in localising the portal density in images of HSV capsids unlike in phages, which have a tail. The PP, pUL6, was well defined in the A-capsid structure located at one fivefold vertex on the outer surface of the capsid as shown by cryo-ET [96]. There is also a strong density within the portal channel and inside of the capsid, which is interpreted as the end of DNA as seen in φ29 and SPP1 [56, 82].

4.4 Tails

The tail organisation in phages depends on their type: *Siphoviridae* have long flexible tails, and *Podoviridae* have very short tails that mostly consist of the adhesive

Phage	Tail proteins	No. of residues	M. Mass kDa	Resolution (Å)	Structure analysis
HK97	putative tail-component IPR010064, gp10	not defined	not known	n/a	None
T5	pb6	464	50	2.2 6	X-ray [97] EM [97]
λ	gpV (TP) gpH (TMP) gpU (terminator)	246 853 131	26 92 15	n/a 2.7	NMR [98, 102] X-ray [103]
SPP1	gp17 (TP) gp17* (TP) gp18 (TMP)	134 264 1032	15 28 111	n/a 14	NMR (gp17) [104] EM [85]
TW1	gp12 (TP) gp14 (TMP)	NF 675	18 72	23	EM [55]
φ29	gp9 (knob) gp12 (tailspike)	599 854	68 92	2.04 1.8–2.05 7.8	X-ray [105, 106] EM [56]
T7	gp11 (TP) gp12 (TP) gp17 (fibres)	196 794 553	22 89 62	12.0 2.0 (X-ray)	EM (gp11,12,17) [87] X-ray (gp17) [107]
P22	gp10 (hub) gp9 (tailspike)	472 667	52 72	9.4 2.0	EM, tomography [88, 91], X-ray [108, 109]
ε15	gp20 (tailspike)	1070	116	20n/r	EM [89, 110]
T4	gp19 (TP) gp15 (terminator)	163 272	18 32	4.11 3.4 15.0 3.2	EM [111, 112] X-ray [113]
HSV	does not have the tail	n/a	n/a	n/a	n/a

Table 3.
Phage tail structures.

device, while *Myoviridae* have rigid long contractile tails that consist of a number of different proteins forming the inner rigid tube and the outer contractile sheath. *Siphoviridae* and *Myoviridae* have an independent pathway for assembly of their tails and are attached to the capsid after it has been packed with genome. However, in *Podoviridae* the tails are assembled on the capsids after DNA packaging as the last step of self-assembly (**Figure 2**, **Table 3**). The long tails of *Siphoviridae* are composed of tail proteins (TPs) that form circular oligomeric rings with three- or sixfold rotational symmetry. The rings are assembled around a tape measure protein (TMP) that defines the length of the tail and are stacked on the top of each other with helical symmetry. A tail terminator protein (TrP) caps the tail when it reaches the length defined by the TMP; the TrP serves as an interface with the capsid. When the phage interacts with the host receptor, the HTI opens and the TMP is pushed out by DNA as a result of the inner pressure of the capsid. Most long tails have a smooth outer surface, but some have appendages that protrude outwards from the tail surface.

T5 phage. The **T5** phage has a long tail (1600 Å) with a diameter of 90 Å. A crystal structure has been obtained for the TP pb6 of T5 (2.2 Å) [97]. The protein consists of two domains: an N-terminal domain with two subdomains both of which comprise a β-sandwich flanked by an α-helix and a long loop (**Figure 6A**) and a C-terminal domain with an immunoglobulin-like fold [98]. The overall tail structure has threefold symmetry (**Figure 6B**) [99, 100]. Cryo-EM was used to

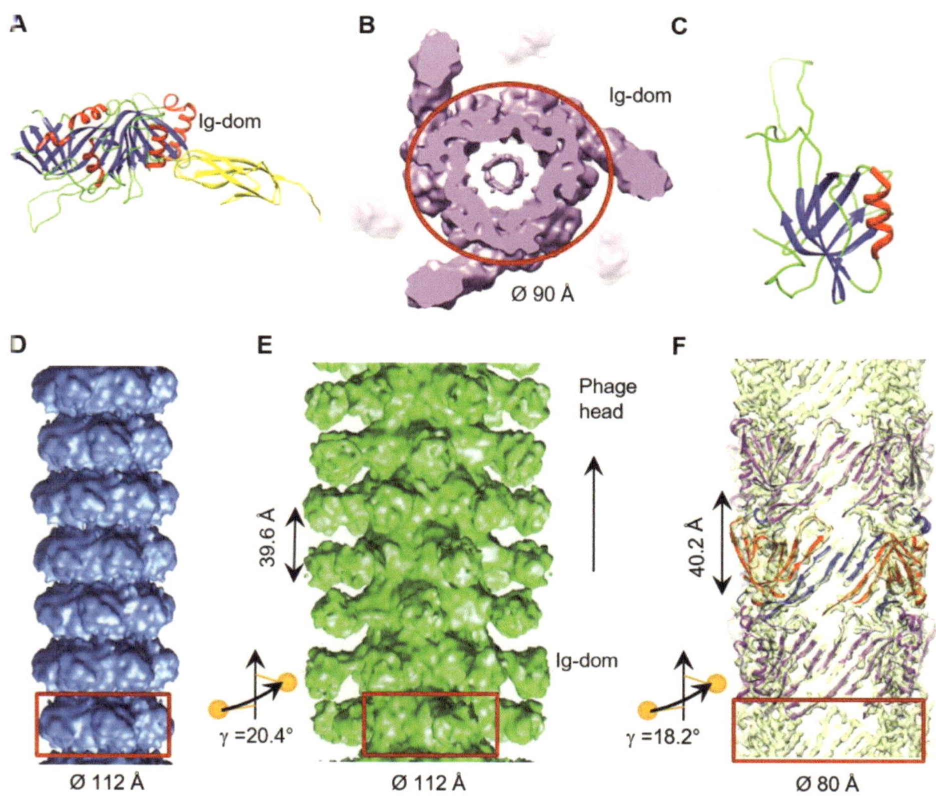

Figure 6.
Bacteriophage tails. (A) The crystal structure of a monomer of T5 pb6 (5NGJ). The extra immunoglobulin domain is coloured yellow. (B) A slice of the combined EM map of T5 (EMD-3692) showing the fold symmetry of the tail. (C) The crystal structure of the N-terminal domain of the P22 TP gpV (2K4Q). (D) Cryo-EM map of SPP1 tails (gp17.1). (E) Cryo-EM map of SPP1 tails (gp17.1). The protrusions are the size of an immunoglobulin domain. (F) The T4 cryo-EM map (EMD-8767) with fitted coordinates (5W5F). Alternate subunits in the central ring are coloured red and blue. The red circle in B and rectangles in D, E and F indicate the inner tail tube, γ—rotation between adjacent tail rings.*

determine the tail structure of T5 at ~6 Å resolution before and after DNA ejection [97] and the atomic model of pb6 used to interpret the results. No differences were found between the two structures, apart from the absence of the tape measure protein pb2 after DNA ejection [97].

λ phage. In the phage λ, the tail has sixfold symmetry and is composed of gpV (TP) and gpH (TMP) [101]. The N-terminal domain of gpV (gpV_N) is required for assembly of the functional phage, and the structure was determined by solution NMR [102]. There are seven β-strands, arranged into two antiparallel sheets which fold into a twisted β-sandwich (**Figure 6C**). The single α-helix is located at the side of the sandwich. The C-terminus of gpV (gpV_C) was shown by solution NMR to have an Ig-like fold [98].

***SPP1** phage*. The structures of the tails from the *Siphoviridae* **SPP1** before and after DNA ejection were determined using negative stain EM at ~14.5 Å resolution [85]. This tail is ~1600 Å long and consists of three proteins gp17 and gp17* that form the tail tube and the TMP gp18. Even at low resolution, the structures revealed movements in the positions of inner domains gp17 and gp17* after DNA ejection [85]. The ratio of proteins gp17 and gp17* within the tail complex suggests that the tail has to have threefold symmetry. Reconstructions of mutants comprising either gp17 or gp17* were obtained and demonstrated sixfold symmetry. A comparison of structures indicated that protein gp17* has an immunoglobulin domain located on the outer surface of the tail (**Figure 6D, E**) (Orlova, Personal communication).

***T4** phage*. *Myoviridae* have contractile tails with a sheath that surrounds a central tail tube. On infection, the tail sheath contracts allowing the tail tube to penetrate the outer membrane of the host cell. The structure of the **T4** phage (a representative of *Myoviridae*) tail is well studied. It consists of a rigid tube, composed of multiple copies of gp19, surrounded by contractile sheath from gp18 subunits [93]. A structure of the central tube at 3.4 Å has been obtained by cryo-EM and showed sixfold symmetry [112] with an axial rise for the helical unit of ~40 Å. This resolution revealed the strands of the β-sheets indicating that four strands from one subunit become part of a continuous helical β-sheet lining the inner channel of the tail (**Figure 6F**). The structure of the T4 contracted tail was obtained at 17 Å resolution, and the tail sheath protein gp18 was found to be arranged into 23 hexameric rings [93]. A crystal structure of a gp18 mutant missing the C-terminal domain [114] was used as the basis for identifying domains in gp18. A homology model based on other *Myoviridae* tail sheath structures [115] was used to localise the C-terminal domain [113].

4.5 Adsorption apparatus

Most *Siphoviridae* phages have an oligomeric ring formed by distal tail proteins (DTPs), which is attached to the last ring of the tail tube [116, 117]. The DTP ring usually serves as an apparatus to recognise and connect to receptor-binding proteins; sometimes, this interaction is assisted by tail fibres found in T4, T5 and other phages. The DTP of SPP1 does not have the fibres [118]. Many phages that infect Gram-negative bacteria have lysozyme-like proteins on the ends of their tails, which enter the periplasm to digest the peptidoglycan barrier [119].

***P22** phage*. *Podoviridae* phage tails have a central needle with trimeric appendages (or spikes) around it, and the part of the tail attached to the capsid could be considered as the baseplate. The short tail of the podovirus **P22** has been studied by cryo-EM of the entire virion [120, 121]. The structure of the P22 tail was obtained at 9.4 Å and aided understanding of its organisation [91]. The 2.8 MDa multisubunit complex is formed by the dodecameric PP gp1, 6 trimeric tailspikes of gp9, 12 copies of the tail accessory protein gp4, a hexon of gp10 and a trimer of the tail needle gp26 (**Figure 7A**). The proteins gp4 and gp10 form a HTI on which the tail assembles and are attached to the PP, while the N-terminal head-binding domain of the outer tailspikes (gp9) attaches to the interface of the gp4 and gp10 subunits. The second larger receptor-binding domain of gp9 contacts and destroys the cell surface lipopolysaccharide [121, 122]. Crystal structures have been obtained for both gp26 [123] and gp9 [108, 109], and these were fitted into the cryo-EM map to find how the different proteins hold the complex together. The tail needle gp26 is folded as triple-stranded coiled coil. The gp9 head-binding domain has a β-helical domain, and the receptor-binding domain is a triangular β-prism (**Figure 7B**) [91]. The asymmetric cryo-EM structure of P22 (7.8 Å) allowed a detailed analysis of the interactions between gp9 and the gp4–gp10 interface by fitting the gp9 and gp26 crystal structures and models of gp1, gp4 and gp10 [92].

***SPP1** phage*. The *Siphoviridae* **SPP1** phage has the long flexible tail. A negative stain reconstruction of the adsorption apparatus revealed a hinge connection between the tip and the tail tube, allowing bending angles as high as 50° [85]. The gp24 protein keeps the tail closed before DNA ejection by forming a cap and is located between the tail and the tip. The first protein in the tip after the cap is gp21,

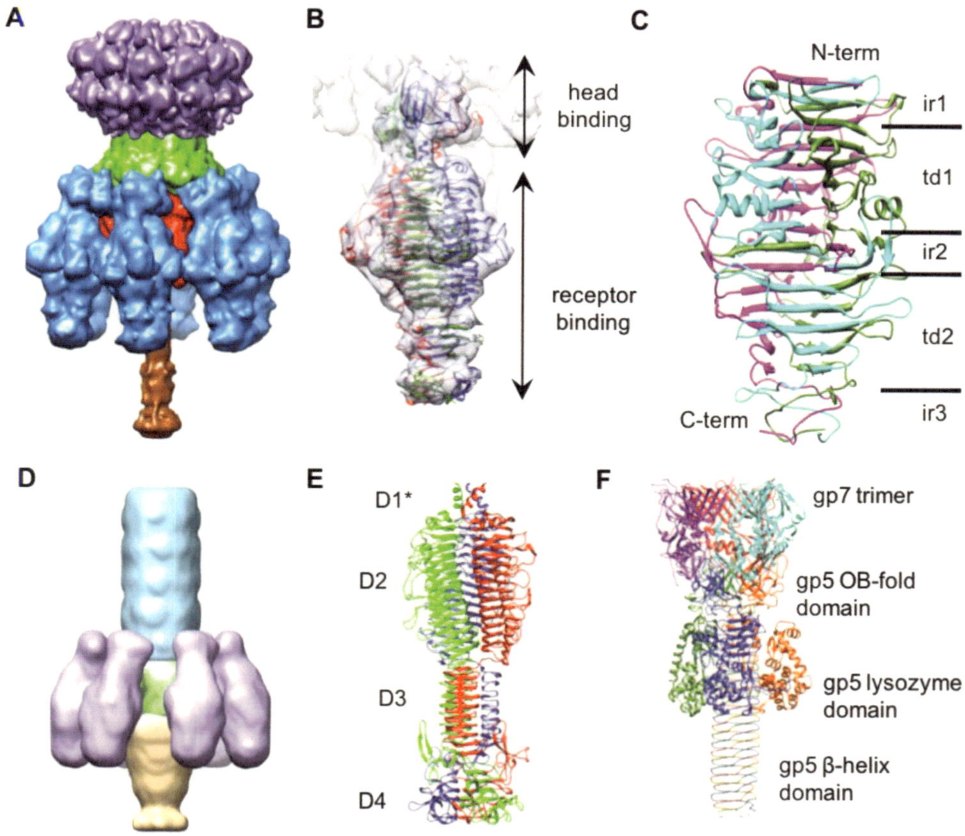

Figure 7.
Adsorption apparatus. (A) The EM map (EMD-5051) of P22 coloured according to the constituent proteins. The PP (gp1), the gp4 12-mer, the gp10 6-mer, the gp9 tailspikes and the gp26 cell-puncturing needle are in purple, green, red, blue and brown, respectively. (B) The crystal structures for the N-terminal head-binding domain (1LKT) and the C-terminal receptor-binding domain (1TYU) of P22 gp9 docked into the cryo-EM density (EMDB-5051). (C) The receptor-binding carboxy-terminal domain of T5 tail fibre pb1 (5AQ5). The five different domain regions are labelled. The N- and C-termini are indicated. (D) The EM map (EMD-8868) of the TW1 tail showing the TP gp12 (in blue), gp15 (in green), gp16, gp18, gp27 (together in light brown) and the gp19 tailspikes (in purple). (E) The φ29 tailspike protein gp12 (3SUC). The four domains D1, D2, D3 and D4 are labelled. (F) The crystal structure (1 K28) of the cell-puncturing device of T4 $(gp27 \pm gp5^* \pm gp5C)_3$. Each monomer within in the gp5 trimer is coloured in blue, green and orange, and each one in gp27 is coloured in magenta, red and cyan.*

and a structural similarity has suggested that SPP1 gp21 has a fold like the P22 tailspike [85]. When the P22 tailspike is fitted to the SPP1 density, the main domain of the P22 tailspike [124] occupies ~70% of the broad flattened area. The most remote protein in the tip is gp19.1, and the predicted secondary structure was structurally similar to the head-binding domain of the P22 tailspike [124]; and fitting three copies of the P22 trimer accounted for all the density in that region. During infection the tip is lost so DNA can pass into the cell and the cap remains in an open state.

***T5** phage*. The crystal structure of the **T5** DTP pb9 has shown that it has two domains. The A-domain has a barrel-like fold with structural similarity to the N-domains of other phage DTPs [118, 125, 126]. In spite of low sequence identity, these proteins form a hexameric ring that occupies the central core of the baseplate. The peripheral B-domain has an oligosaccharide-/oligonucleotide-binding (OB) fold [127]. The attachment of phage T5 to the host cell is assisted by three side tail fibres attached to the distal end of the tail [107, 127, 128], and they are homo-trimers of the pb1 (1396 aa). The trimeric structure of the receptor-binding carboxy-terminal domain 970–1263(aa) was determined at 2.3 Å using X-ray crystallography [107] and could be divided into five different regions (**Figure 7C**) based on the structure of the P22 spike [124]. The N-terminal region (989-1009 aa) is shaped by the β-strands of the three monomers that wrap around each other to form a threefold beta-helix [124, 129]. The first "interdigitated" region (**ir1**) is followed by a triangular domain (1010–1129 aa) where three concave β sheets form a β-prism (**td1**). The second interdigitated region **ir2** (1130–1160 aa) also forms a short triple beta-helix. A second triangular domain **td2** (1161–1238 aa) is a β-prism like **td1**. At the distal end of the fibre, the third interdigitated region (aa1239–1263), **ir3**, forms a tapered triple-helical structure making the end of the structure pointed (**Figure 7C**). There is some similarity in the structure with the P22 tailspike [124] as both have a β-helical domain, an **ir** region, a triangular beta-prism domain and a second **ir** domain (called caudal fin). The triangular β-prism of P22 is the most similar to **td2** of pb1 and has the same topology.

TW1** phage*. ***TW1 has an unusual tail organisation for a siphophage, as a cryo-EM reconstruction of the tail (23.5 Å) [55] revealed six spikes on the distal end from the head. They are attached to the central tail tube, similar to the spikes seen in podophages P22 and Sf6 [120, 130] (**Figure 7D**). The TW1 gp19 tailspike (TS) protein is homologous to the TS protein of the podophage HK620 [131] so the crystal structure of the HK620 TS protein was fitted into the TW1 appendages. The TW1 gp19 TSs are thought to be attached to the phage via the DTP gp15 protein. However, the size of TW1 gp15 and the EM density suggest that this protein does not have a peripheral OB-fold domain as seen in the DTP of phage T5 [127]. Below gp15 are gp16 and gp18, which form the central tip of the phage tail (**Figure 7D**) and are similar to phage λ proteins gpL and gpJ, respectively [132]. At the tip of the tail is gp27 (**Figure 7D**) which is homologous to peptidoglycan-degrading enzymes. Many phages that infect Gram-negative bacteria have lysozyme-like proteins in their tails which enter the periplasm to digest the peptidoglycan barrier [119].

φ29** phage*. The tail of ***φ29 has 12 appendages, which are similar to the tailspikes of phage P22 and are attached to the bulge of densities close to the capsid [133]. Each appendage is a trimer of gp12* (the cleavage product of gp12 which during maturation loses an 18 kDa C-terminal fragment). Although there is no sequence similarity with the P22 tailspike, the P22 tailspike domain structure gave a good fit into the peripheral component of the φ29 appendages [65, 124]. A construct of gp12 residues 89–854 was cleaved in vivo to give an N-terminal fragment (up to Ser691) and a C-terminal fragment (from Asp692) and crystal structures obtained for each [105]. They are both trimers, and the N-terminal part attaches to the virion and has three domains: D1* is a coiled coil, D2 is mostly a β-helix and D3 is also a β-helix (**Figure 7E**). The C-terminal domain D4 acts as a chaperone for trimer assembly and is cleaved by autocatalysis. The φ29 structure attached to the lipid bilayer has been obtained by cryo-ET (34 Å) [134]. The structure is comparable to cryo-EM structures of mature φ29 [56, 133, 135]. Tomographic reconstructions demonstrated the different stages of infection [134]. In the adsorption stage, the phage is tilted to the cell wall, and both the appendages and the tail seem to contact the cell surface. The tail tip protein helps the phage penetrate the cell wall. When it contacts the cytoplasmic membrane, a pore is created which allows the genome to be injected into the cell.

***T4** phage*. The structure of the baseplate in *Myoviridae* is complex as illustrated by the **T4** phage. The sixfold symmetric baseplate is 270 Å long and about 520 Å in diameter at the base and is connected to the distal end of the tail [136, 137]. It is composed at least by 16 different proteins [137]. A star-shaped baseplate is formed by sequential binding of four different proteins to form a wedge shape [137]. Six wedges are arranged around the independently assembled hub. Finally, other proteins are added to form the complete baseplate. Once gp48 and gp54 have bound to the top of the central hub, polymerisation of the tail tube is initiated, and after gp25 has attached to them, then polymerisation of the sheath is initiated [138]. Crystal structures of these constituent proteins were fitted into EM structure, and this showed the location of the proteins [137]. Six long fibres and six short fibres are attached to the baseplate. The long fibres reversibly interact with the cell surface receptors [139]. After recognition, the baseplate comes closer to the cell surface allowing the six short tail fibres to bind irreversibly to the cell outer membrane. This process is accompanied by a large conformational change in the baseplate from a "high-energy" to a "low-energy" structure [93, 140]. This induces contraction of the tail sheath and allows the inner tail tube to pierce the outer host cell membrane and penetrate the inner membrane so that the genome is transferred directly to the host's cytoplasm.

The structure of the T4 baseplate was assembled in vitro from gp10, gp7, gp8, gp6 and gp53, and the crystal structure was determined (4.2 Å) [141]. This indicated interesting differences compared to the structures when they are separately crystallised. However, about two-thirds of the structure was missing, but a cryo-EM structure of the same construct (3.8 Å) provided the positions of these missing parts [142]. The structures of T4 baseplate in its pre- and post-host attachment states were determined at 4.11 and 6.77 Å, respectively, by cryo-EM [111]. By combining high-resolution structures of the individual baseplate proteins, the authors were able to build a pseudo-atomic model for the baseplate proteins. The crystal structure at 2.9 Å of the gp5–gp27 cell-puncturing device was fitted into the EM structure (**Figure 7F**) [143]. Positions of gp27, gp5C (the C-terminal β-helix domain of gp5) and gp5* (the N-terminal OB-fold domain and the lysozyme middle domain) were identified. A monomeric protein gp5.4 caps the tip of the gp5 β-helix to sharpen the central spike [144]. During infection this spike punctures the cell membrane, and the lysozyme domain of gp5 digests the peptidoglycan in the *E. coli* periplasm.

***ε15** phage.* A 20 Å cryo-EM map of **ε15** showed six gp20 tailspikes extending out from one of the fivefold capsid vertices. Each tailspike is composed of two domains [89] and has slightly different orientations with respect to the capsid. Cryo-ET has been used to show the interaction of ε15 phage with the cell and to visualise the process of how ε15 infects its host *Salmonella anatum* [110]. Initially, the tailspikes attach to the host cell followed by the tail hub attaching to a putative cell receptor. A bowl-shaped density was observed beneath the tail hub at the beginning of infection. It was proposed that phage indents the host outer membrane looking for a secondary receptor or for puncturing the membrane. A tunnel is established through the cell wall which allows the DNA to enter the cell.

5. Conclusions

Structural studies of the currently known tailed phages have shown a common organisation, which implies that they have a single ancestor and diversity has arisen through evolution [37]. All phages have a similar pathway of self-assembly: a procapsid formed with the help of a SP (or sometimes a scaffolding domain); conformational changes induced by release of the SP create a space for the DNA, and assisted by DNA terminases, the genome is packaged into the procapsid. This step is typically named as the maturation of the capsid. The tail is then attached or assembled on the capsid to form the infectious virion. The MCPs are characterised by the HK97 capsid protein fold. However, phages have a very low sequence similarity, which leads to differences in how the capsid stability is arranged to withstand the high inner pressure of the genome. In some phages like HK97 and SPP1, the interactions between capsid proteins are strong and hold the capsid intact. In many phages the process of capsid maturation is linked to attachment of additional proteins that are named as auxiliary or decoration proteins. They are often essential to enhance the capsid stability. The HK97 capsid is held together by chain mail covalent links between the MCPs; in SPP1 and T5, the decoration proteins enhance stability of the capsid, but in λ, T4 and ε15 phages, these proteins are essential for keeping DNA inside the capsid [19, 52, 53].

The HTIs play an important role in all tailed phages as they provide a channel for DNA to enter and exit the capsid and at the same time provide a covalent connection to either the preassembled tails or tails assembled on the capsid. They all contain a dodecameric PP positioned within the capsid at one of the fivefold vertices and that acts as a gatekeeper holding the DNA within the capsid even in very harsh environments. Like the capsid proteins, the PPs have a common fold with the conserved elements being involved in interactions with DNA [145]. They have mostly α-helical domains in their central part and β-layers in the wing domains that interact with the capsid to fix the PP position. Head completion proteins below the PP also have similar folds to each other.

A much higher level of divergence is reflected in phage tail structures. The most common feature in all long-tailed phages is a central tube with a large number (30–40) of three- or sixfold circular rings of the major TPs. There is structural similarity between these major TPs: they have a similar fold of a β-sandwich flanked by alpha-helices and loops that provide links between adjacent rings. The helical tails have a typical rise of about 40 Å and rotation of around 20° between adjacent rings. Some tails also have appendages, which appear to have an immunoglobulin-like fold. Very little is known about the organisation of tail sheaths that have some similarities with type VI secretion systems, but sometimes they have extra appendages like immunoglobulin domains to help phages recognise their host cells. There is also some structural similarity of the TP with the tail terminator proteins and proteins in the T4 sheath.

Even higher diversity is found in the adsorption apparatus which are responsible for the recognition of the host cells and signalling the opening of the gate for the genome release. The tip of phage SPP1 recognises its receptor; induces the tail to be attached to the outer membrane of the host cell after disconnection of the tip. At the same time this interaction generates a signal that open the PP gate keeper. The T4 phage has a significantly more complex system of a baseplate which undergoes several steps of complex conformational changes.

Interestingly, the receptor-binding proteins also a have similar organisation: they are all trimers, usually intertwined with β-helical regions, and use their N-terminal domain to bind to the phage. Spikes and fibres are also found in many phages. However, the number of spikes or fibres varies significantly between phages. Podophages have trimeric tailspikes to recognise the specific host cell for infection. Like other phage components, they vary from six fibres in phage T7 to 12 in phage φ29, but they all have a β-helical fold. The fibres can have different roles within a phage, for instance, T4 has six long fibres that serve as host recognition and six short fibres which then extend and bind to the cell.

Antibiotics (especially of the broad-spectrum type) are very effective at killing infectious bacteria; however, they kill typically multiple bacterial species indiscriminately, thus destroying beneficial bacteria of the host microbiome as well. Since phages are specific to one species of bacteria, they are unlikely to perturb microbiome bacterial species. Current problems with antibiotic resistance require new approaches, and here phages can be used [12]. For medicinal purposes it is necessary to design a phage that will recognise the specific bacteria we want to eliminate [146]. Phages can be modified for high specificity in the recognition of pathogens. The high level of phage specificity is based on recognition of receptor characteristic for a given type of bacteria which is where the differences in the adsorption systems of different phages play a crucial role. The important task in studying phages is to find those that are able to kill only antibiotic-resistant bacteria. Here, the lytic phages are of most interest, since rather than stopping bacteria from producing a certain type of protein that will slow down the bacterium proliferation, like in the case of antibiotics, these phages destroy the bacteria's cell wall and cell membrane completely. In addition, many bacteria develop biofilm—a thick layer of viscous materials that protect them from antibiotics. Some phages are equipped with tools that can digest this biofilm [147]. There are some problems with phages, since they are easy to use for topical applications, but often specific medications have to be administered internally. For phages to be used for delivery of drugs, they need to be more precise in their action. Consequently, we need to modify them so that the infectivity will be efficient by replacing the genome with DNA encoding specific enzymes and the adsorption apparatus made more effective. To develop these medical approaches, we need to know the phage organisation and the interactions between protein components at the atomic level. To achieve this hybrid, methods should be used including structural biology, biochemistry and microbiology [21].

Acknowledgements

The authors are grateful to Dr. D. Houldershaw and Mr. Y. Goudetsidis for their computer support. The authors apologise for the incomplete coverage of known phage structures and have drawn on a limited subset, owing to space constraints.

Conflict of interest

The authors declare no conflict of interest.

Author details

Helen E. White and Elena V. Orlova*
Institute of Structural and Molecular Biology, Birkbeck College, London, United Kingdom

*Address all correspondence to: e.orlova@mail.cryst.bbk.ac.uk

References

[1] Shors T. Understanding Viruses. Sudbury, USA: Jones and Bartlett Publishers; 2008. ISBN: 0-7637-2932-9

[2] Brussow H, Hendrix RW. Phage genomics: Small is beautiful. Cell. 2002;**108**:13-16

[3] Suttle CA. Marine viruses—Major players in the global ecosystem. Nature Reviews. Microbiology. 2007;**5**:801-812

[4] Breitbart M. Marine viruses: Truth or dare. Annual Review of Marine Science. 2012;**4**:425-448

[5] Coelho J, Woodford N, Turton J, Livermore DM. Multiresistant *Acinetobacter* in the UK: How big a threat? Journal of Hospital Infection. 2004;**58**:167-169

[6] Hanlon GW. Bacteriophages: An appraisal of their role in the treatment of bacterial infections. International Journal of Antimicrobial Agents. 2007;**30**:118-128

[7] Burrowes B, Harper DR, Anderson J, McConville M, Enright MC. Bacteriophage therapy: Potential uses in the control of antibiotic-resistant pathogens. Expert Review of Anti-Infective Therapy. 2011;**9**:775-785

[8] Coates AR, Halls G, Hu Y. Novel classes of antibiotics or more of the same? British Journal of Pharmacology. 2011;**163**(1):184-194. DOI: 10.1111/j.1476-5381.2011.01250.x

[9] Brussow H, Kutter E. Phage ecology. In: Kutter E, Sulakvelidze A, editors. Bacteriophages: Biology and applications. Boca Raton, FL: CRC Press; 2005. pp. 129-163

[10] Malik DJ, Sokolov IJ, Vinner GK, Mancuso F, Cinquerrui S, Vladisavljevic GT, et al. Formulation, stabilisation and encapsulation of bacteriophage for phage therapy. Advances in Colloid and Interface Science. 2017;**249**:100-133

[11] Weinbauer MG. Ecology of prokaryotic viruses. FEMS Microbiology Reviews. 2004;**28**:127-181

[12] Criscuolo E, Spadini S, Lamanna J, Ferro M, Burioni R. Bacteriophages and their immunological applications against infectious threats. Journal of Immunology Research. 2017;**2017**:3780697. Published online: Apr 6, 2017. DOI: 10.1155/2017/3780697

[13] Pietilä MK, Demina TA, Atanasova NS, Oksanen HM, Bamford DH. Archaeal viruses and bacteriophages: Comparisons and contrasts. Trends in Microbiology . 2014;**22**:6334-6344

[14] Orlova EV. Bacteriophages and Their Structural Organisation. In: Kurtböke İ, editor. Bacteriophages. IntechOpen; 2012. DOI: 10.5772/1065. ISBN: 978-953-51-0272-4

[15] Ackermann HW. Classification of bacteriophages. In: Calendar R, editor. The Bacteriophages. New York, USA: Oxford University Press; 2006. pp. 8-16. ISBN: 0-19-514850-9

[16] Drenth J. Principles of Protein X-Ray Crystallography. New York: Springer-Verlag; 2007

[17] Javed A, Christodoulou J, Cabrita LD, Orlova EV. The ribosome and its role in protein folding: Looking through a magnifying glass. Acta Crystallographica. Section D, Structural Biology. 2017;**73**(Pt 6):509-521. DOI: 10.1107/S2059798317007446

[18] Grimes JM, Burroughs JN, Gouet P, Diprose JM, Malby R, Ziéntara S, et al. The atomic structure of the bluetongue virus core. Nature. 1998;**395**:470-478

[19] Wikoff WR, Liljas L, Duda RL, Tsuruta H, Hendrix RW, Johnson JE. Topologically linked protein rings in the bacteriophage HK97 capsid. Science. 2000;**289**:2129-2133

[20] Cuniasse P, Tavares P, Orlova EV, Zinn-Justin S. Structures of biomolecular complexes by combination of NMR and cryoEM methods. Current Opinion in Structural Biology. 2017;**43**:104-113. DOI: 10.1016/j.sbi.2016.12.008

[21] Lengyel J, Hnath E, Storms M, Wohlfarth T. Towards an integrative structural biology approach: combining Cryo-TEM, X-ray crystallography, and NMR. Journal of Structural and Functional Genomics. 2014;**15**:117-124

[22] Ram S, Ward ES, Ober RJ. Beyond Rayleigh's criterion: A resolution measure with application to single-molecule microscopy. Proceedings of the National Academy of Sciences of the United States of America. 2006;**103**(12):4457-4462

[23] Orlova EV, Saibil HR. Structural analysis of macromolecular assemblies by electron microscopy. Chemical Reviews. 2011;**111**(12):7710-7748. DOI: 10.1021/cr100353t

[24] Harris JR. Negative Staining and Cryoelectron Microscopy, The Thin Film Techniques. Oxford, UK: BIOS Scientific Publishers; 1997. ISBN: 1859961207

[25] Adrian M, Dubochet J, Lepault J, McDowall AW. Cryo-electron microscopy of viruses. Nature. 1984;**308**(5954):32-36

[26] Ruskin RS, Yu Z, Grigorieff N. Quantitative characterization of electron detectors for transmission electron microscopy. Journal of Structural Biology. Journal of Structural Biology. 2013;**184**(3):385-393. DOI: 10.1016/j.jsb.2013.10.016. Published online: Nov 1, 2013, 10.1016/j.jsb.2013.10.016

[27] Bartesaghi A, Aguerrebere C, Falconieri V, Banerjee S, Earl LA, Zhu X, et al. Atomic resolution Cryo-EM structure of β-Galactosidase. Structure. 2018;**26**(6):848-856.e3. DOI: 10.1016/j.str.2018.04.004

[28] Wan W, Briggs JA. Cryo-electron tomography and subtomogram averaging. Methods in Enzymology. 2016;**579**:329-367. DOI: 10.1016/bs.mie.2016.04.014

[29] Briggs JAG. Structural biology in situ—The potential of subtomogram averaging. Current Opinion in Structural Biology. 2013;**23**:261-267

[30] Tavares P. The bacteriophage head-to-tail interface. Sub-Cellular Biochemistry. 2018;**88**:305-328. DOI: 10.1007/978-981-10-8456-0_14

[31] Orlova EV, Dube P, Beckmann E, Zemlin F, Lurz R, Trautner TA, et al. Structure of the 13-fold symmetric portal protein of bacteriophage SPP1. Nature Structural Biology. 1999;**6**:842-846

[32] Cingolani G, Moore SD, Prevelige PE Jr, Johnson JE. Preliminary crystallographic analysis of the bacteriophage P22 portal protein. Journal of Structural Biology. 2002;**139**:46-54

[33] Cerritelli ME, Trus BL, Smith CS, Cheng N, Conway JF, Steven AC. A second symmetry mismatch at the portal vertex of bacteriophage T7: 8-fold symmetry in the procapsid core. Journal of Molecular Biology. 2003;**327**:1-6

[34] Trus BL, Cheng N, Newcomb WW, Homa FL, Brown JC, Steven AC. Structure and polymorphism of the UL6 portal protein of herpes simplex virus type 1. Journal of Virology. 2004;**78**:12668-12671

[35] Lorenzen K, Olia AS, Uetrecht C, Cingolani G, Heck AJ. Determination of stoichiometry and conformational changes in the first step of the P22 tail assembly. Journal of Molecular Biology. 2008;**379**:385-396

[36] Hendrix RW, Smith MCM, Burns RN, Ford ME, Hatfull GF. Evolutionary relationships among diverse bacteriophages and prophages: All the world's a phage. Proceedings of the National Academy of Sciences of the United States of America. 1999;**96**:2192-2197

[37] Bamford DH, Grimes JM, Stuart DI. What does structure tell us about virus evolution? Current Opinion in Structural Biology. 2005;**15**:655-663

[38] Wikoff WR, Conway JF, Tang J, Lee KK, Gan L, Cheng N, et al. Time-resolved molecular dynamics of bacteriophage HK97 capsid maturation interpreted by electron cryo-microscopy and X-ray crystallography. Journal of Structural Biology. 2006;**153**(3):300-306

[39] Huet A, Conway JF, Letellier L, Boulanger P. In vitro assembly of the T = 13 procapsid of bacteriophage T5 with its scaffolding domain. Journal of Virology. 2010;**84**:9350-9358. DOI: 10.1128/JVI.00942-10

[40] Veesler D, Quispe J, Grigorieff N, Potter CS, Carragher B, Johnson JE. Maturation in action: CryoEM study of a viral capsid caught during expansion. Structure. 2012;**20**:1384-1390

[41] Baker TS, Olson NH, Fuller SD. Adding the third dimension to virus life cycles: Three-dimensional reconstruction of icosahedral viruses from cryo-electron micrographs. Microbiology and Molecular Biology Reviews. 1999;**63**(4):862-922. (Review)

[42] Helgstrand C, Wikoff WR, Duda RL, Hendrix RW, Johnson JE, Liljas L. The refined structure of a protein catenane: the HK97 bacteriophage capsid at 3.44 Å resolution. Journal of Molecular Biology. 2003;**334**:885-899

[43] Atanasova NS, Bamford DH, Oksanen HM. Haloarchaeal virus morphotypes. Biochimie. 2015;**118**:333-343

[44] Baker ML, Jiang W, Rixon FJ, Chiu W. Common ancestry of herpesviruses and tailed DNA bacteriophages. Journal of Virology. 2005;**79**:14967-14970

[45] Huet A, Makhov AM, Huffman JB, Vos M, Homa FL, Conway JF. Extensive subunit contacts underpin herpesvirus capsid stability and interior-to-exterior allostery. Nature Structural & Molecular Biology. 2016;**23**:531-539

[46] Yu X, Jih J, Jiang J, Hong Zhou Z. Atomic structure of the human cytomegalovirus capsid with its securing tegument layer of pp150. Science. 2017;**356**:1350

[47] Lander GC, Evilevitch A, Jeembaeva M, Potter CS, Carragher B, Johnson JE. Bacteriophage lambda stabilization by auxiliary protein gpD: Timing, location, and mechanism of attachment determined by cryoEM. Structure. 2008;**16**(9):1399-1406. DOI: 10.1016/j.str.2008.05.016

[48] Yang F, Forrer P, Dauter Z, Conway JF, Cheng N, Cerritelli ME, et al. Novel fold and capsid-binding properties of the lambda-phage display platform protein gpD. Nature Structural Biology. 2000;**7**(3):230-237

[49] Jiang W, Baker ML, Jakana J, Weigele PR, King J, Chiu W. Backbone structure of the infectious epsilon15 virus capsid revealed by electron cryomicroscopy. Nature. 2008;**451**(7182):1130-1134

[50] Baker ML, Hryc CF, Zhang Q, Wu W, Jakana J, Haase-Pettingell C, et al. Validated near-atomic resolution structure of bacteriophage epsilon15 derived from cryo-EM and modeling.

Proceedings of the National Academy of Sciences of the United States of America. 2013;**110**:12301-12306

[51] Conway JF, Wikoff WR, Cheng N, Duda RL, Hendrix RW, Johnson JE, et al. Virus maturation involving large subunit rotations and local refolding. Science. 2001;**292**:744-748

[52] Vernhes E, Renouard M, Gilquin B, Cuniasse P, Durand D, England P, et al. High affinity anchoring of the decoration protein pb10 onto the bacteriophage T5 capsid. Scientific Reports. 2017;**7**:41662. DOI: 10.1038/srep41662

[53] White HE, Sherman MB, Brasilès S, Jacquet E, Seavers P, Tavares P, et al. Capsid structure and its stability at the late stages of bacteriophage SPP1 assembly. Journal of Virology. 2012;**12**:6768-6777. DOI: 10.1128/JVI.00412-12

[54] Bebeacua C, Lai L, Vegge CS, Brøndsted L, van Heel M, Veesler D, et al. Visualizing a complete Siphoviridae member by single-particle electron microscopy: The structure of lactococcal phage TP901-1. Journal of Virology. 2013;**87**(2):1061-1068. DOI: 10.1128/JVI.02836-12

[55] Wang Z, Hardies SC, Fokine A, Klose T, Jiang W, Cho BC, et al. Structure of the marine siphovirus TW1: Evolution of capsid-stabilizing proteins and tail spikes. Structure. 2018;**26**:238-248

[56] Tang J, Olson N, Jardine PJ, Grimes S, Anderson DL, Baker TS. DNA poised for release in bacteriophage phi29. Structure. 2008;**16**:935-943

[57] Guo F, Liu Z, Fang PA, Zhang Q, Wright ET, Wu W, et al. Capsid expansion mechanism of bacteriophage T7 revealed by multistate atomic models derived from cryo-EM reconstructions. Proceedings of the National Academy of Sciences of the United States of America. 2014;**111**(43):E4606-E4614. DOI: 10.1073/pnas.1407020111

[58] Parent KN, Khayat R, Tu LH, Suhanovsky MM, Cortines JR, Teschke CM, et al. P22 coat protein structures reveal a novel mechanism for capsid maturation: stability without auxiliary proteins or chemical crosslinks. Structure. 2010;**18**(3):390-401. DOI: 10.1016/j.str.2009.12.014

[59] Chen DH, Baker ML, Hryc CF, DiMaio F, Jakana J, Wu W, et al. Structural basis for scaffolding-mediated assembly and maturation of a dsDNA virus. Proceedings of the National Academy of Sciences of the United States of America. 2011;**108**(4):1355-1360

[60] Hryc CF, Chen DH, Afonine PV, Jakana J, Wang Z, Haase-Pettingell C, et al. Accurate model annotation of a near-atomic resolution cryo-EM map. Proceedings of the National Academy of Sciences of the United States of America. 2017;**114**:3103-3108

[61] Fokine A, Leiman PG, Shneider MM, Ahvazi B, Boeshans KM, Steven AC, et al. Structural and functional similarities between the capsid proteins of bacteriophages T4 and HK97 point to a common ancestry. Proceedings of the National Academy of Sciences of the United States of America. 2005;**102**(20):7163-7168

[62] Chen Z, Sun L, Zhang Z, Fokine A, Padilla-Sanchez V, Hanein D, et al. Cryo-EM structure of the bacteriophage T4 isometric head at 3.3-angstrom resolution and its relevance to the assembly of icosahedral viruses. Proceedings of the National Academy of Sciences of the United States of America. 2017;**114**:E8184-E8193

[63] Dai X, Zhou ZH. Structure of the herpes simplex virus 1 capsid with associated tegument protein complexes. Science. 2018;**360**(6384. pii: eaao7298). DOI: 10.1126/science.aao7298

[64] Holden HM, Ito M, Hartshorne DJ, Rayment I. X-ray structure determination

of telokin, the C-terminal domain of myosin light chain kinase, at 2.8 Å resolution. Journal of Molecular Biology. 1992;**227**:840-851

[65] Xiang Y, Morais MC, Battisti AJ, Grimes S, Jardine PJ, Anderson DL, et al. Structural changes of bacteriophage φ29 upon DNA packaging and release. The EMBO Journal. 2006;**25**:5229-5239

[66] Morais MC, Choi KH, Koti JS, Chipman PR, Anderson DL, Rossmann MG. Conservation of the capsid structure in tailed dsDNA bacteriophages: The pseudoatomic structure of φ29. Molecular Cell. 2005;**18**:149-159

[67] Marchler-Bauer A, Anderson JB, Cherukuri PF, DeWeese-Scott C, Geer LY, Gwadz M, et al. CDD: A conserved domain database for protein classification. Nucleic Acids Research. 2005;**33**:D192-D196

[68] Iwasaki K, Trus BL, Wingfield PT, Cheng N, Campusano G, Rao VB, et al. Molecular architecture of bacteriophage T4 capsid: Vertex structure and bimodal binding of the stabilizing accessory protein, Soc. Journal of Virology. 2000;**271**:321-333

[69] Sun L, Zhang X, Gao S, Rao PA, Padilla-Sanchez V, Chen Z, et al. Cryo-EM structure of the bacteriophage T4 portal protein assembly at near-atomic resolution. Nature Communications. 2015;**6**:7548

[70] Fokine A, Chipman PR, Leiman PG, Mesyanzhinov VV, Rao VB, Rossmann MG. Molecular architecture of the prolate head of bacteriophage T4. Proceedings of the National Academy of Sciences of the United States of America. 2004;**101**:6003-6008

[71] Fokine A, Islam MZ, Zhang Z, Bowman VD, Rao VB, Rossmann MG. Structure of the three N-terminal immunoglobulin domains of the highly immunogenic outer capsid protein from a T4-like bacteriophage. Journal of Virology. 2011;**85**:8141-8148

[72] Qin L, Fokine A, O'Donnell E, Rao VB, Rossmann MG. Structure of the small outer capsid protein, Soc: A clamp for stabilizing capsids of T4-like phages. Journal of Molecular Biology. 2010;**395**(4):728-741. DOI: 10.1016/j.jmb.2009.10.007

[73] Heymann JB, Cheng N, Newcomb WW, Trus BL, Brown JC, Steven AC. Dynamics of herpes simplex virus capsid maturation visualized by time-lapse cryo-electron microscopy. Nature Structural Biology. May 2003;**10**(5):334-341

[74] Cardarelli L, Lam R, Tuite A, Baker LA, Sadowski PD, Radford DR, et al. The crystal structure of bacteriophage HK97 gp6: defining a large family of head-tail connector proteins. Journal of Molecular Biology. 2010;**395**:754-768

[75] Simpson AA, Tao Y, Leiman PG, Badasso MO, He Y, Jardine PJ, et al. Structure of the bacteriophage ϕ29 DNA packaging motor. Nature. 2000;**408**:745-750

[76] Guasch A, Pous J, Ibarra B, Gomis-Ruth FX, Valpuesta JM, Sousa N, et al. Detailed architecture of a DNA translocating machine: The high-resolution structure of the bacteriophage phi29 connector particle. Journal of Molecular Biology. 2002;**315**:663-676

[77] Lebedev AA, Krause MH, Isidro AL, Vagin A, Orlova EV, Turner J, et al. Structural framework for DNA translocation via the viral portal protein. The EMBO Journal. 2007;**26**:1984-1994

[78] Olia AS, Prevelige PE Jr, Johnson JE, Cingolani G. Three-dimensional structure of a viral genome-delivery portal vertex. Nature Structural & Molecular Biology. 2011;**18**:597-603

[79] Agirrezabala X, Martín-Benito J, Valle M, González JM, Valencia A, Valpuesta JM, et al. Structure of the connector of bacteriophage T7 at 8 Å resolution: Structural homologies of a basic component of a DNA translocating machinery. Journal of Molecular Biology. 2005;**347**:895-902

[80] Sun S, Kondabagil K, Draper B, Alam TI, Bowman VD, Zhang Z, et al. The structure of the phage T4 DNA packaging motor suggests a mechanism dependent on electrostatic forces. Cell. 2008;**135**:1251-1262

[81] Oliveira L, Tavares P, Alonso JC. Headful DNA packaging: Bacteriophage SPP1 as a model system. Virus Research. 2013;**173**(2):247-259. DOI: 10.1016/j.virusres.2013.01.021

[82] Chaban Y, Lurz R, Brasiles S, Cornilleau C, Karreman M, Zinn-Justin S, et al. Structural rearrangements in the phage head-to-tail interface during assembly and infection. Proceedings of the National Academy of Sciences of the United States of America. 2015;**112**:7009-7014

[83] Orlova EV, Gowen B, Dröge A, Stiege A, Weise F, Lurz R, et al. Structure of a viral DNA gatekeeper at 10 A resolution by cryoelectron microscopy. The EMBO Journal. 2003;**22**(6):1255-1262

[84] Lhuillier S, Gallopin M, Gilquin B, Brasilès S, Lancelot N, Letellier G, et al. Structure of bacteriophage SPP1 head-to-tail connection reveals mechanism for viral DNA gating. Proceedings of the National Academy of Sciences of the United States of America. 2009;**106**(21):8507-8512

[85] Plisson C, White HE, Auzat I, Zafarani A, São-Jose C, Lhuillier S, et al. Structure of bacteriophage SPP1 tail reveals trigger for DNA ejection. The EMBO Journal. 2007;**26**:3720-3728

[86] Auzat I, Dröge A, Weise F, Lurz R, Tavares P. Origin and function of the two major tail proteins of bacteriophage SPP1. Molecular Microbiology. 2008 Nov;**70**(3):557-569. DOI: 10.1111/j.1365-2958.2008.06435.x

[87] Cuervo A, Pulido-Cid M, Chagoyen M, Arranz R, González-García VA, Garcia-Doval C, et al. Structural characterization of the bacteriophage T7 tail machinery. The Journal of Biological Chemistry. 2013;**288**:26290-26299

[88] Pintilie G, Chen DH, Haase-Pettingell CA, King JA, Chiu W. Resolution and probabilistic models of components in CryoEM maps of mature P22 bacteriophage. Biophysical Journal. 2016;**110**(4):827-839. DOI: 10.1016/j.bpj.2015.11.3522

[89] Jiang W, Chang J, Jakana J, Weigele P, King J, Chiu W. Structure of epsilon15 bacteriophage reveals genome organization and DNA packaging/injection apparatus. Nature. 2006;**439**:612-616

[90] McElwee M, Vijayakrishnan S, Rixon F, Bhella D. Structure of the herpes simplex virus portal-vertex. PLoS Biology. 2018;**16**(6):e2006191. DOI: 10.1371/journal.pbio.2006191

[91] Lander GC, Khayat R, Li R, Prevelige PE, Potter CS, Carragher B, et al. The P22 tail machine at subnanometer resolution. Structure. 2009;**17**(6):789-799

[92] Tang J, Lander GC, Olia A, Li R, Casjens S, Prevelige P Jr, et al. Peering down the barrel of a bacteriophage portal: The genome packaging and release valve in p22. Structure. 2011;**19**:496-502

[93] Leiman PG, Chipman PR, Kostyuchenko VA, Mesyanzhinov VV, Rossmann MG. Three-dimensional rearrangement of proteins in the tail

of bacteriophage T4 on infection of its host. Cell. 2004;**118**:419-429

[94] Isidro A, Santos MA, Henriques AO, Tavares P. The high-resolution functional map of bacteriophage SPP1 portal protein. Molecular Microbiology. 2004;**51**:949-962

[95] Newcomb WW, Juhas RM, Thomsen DR, Homa FL, Burch AD, Weller SK, et al. The UL6 gene product forms the portal for entry of DNA into the herpes simplex virus capsid. Journal of Virology. 2001;**75**:10923-10932

[96] Cardone G, Winkler DC, Trus BL, Cheng N, Heuser JE, Newcomb WW, et al. Visualization of the herpes simplex virus portal in situ by cryo-electron tomography. Virology. 2007;**361**:426-434

[97] Arnaud CA, Effantin G, Vives C, Engilberge S, Bacia M, Boulanger P, et al. Bacteriophage T5 tail tube structure suggests a trigger mechanism for Siphoviridae DNA ejection. Nature Communications. 2017;**8**:1953-1953

[98] Pell LG, Gasmi-Seabrook GM, Morais M, Neudecker P, Kanelis V, Bona D, et al. The solution structure of the C-terminal Ig-like domain of the bacteriophage λ tail tube protein. Journal of Molecular Biology. 2010;**403**:468-479

[99] Effantin G, Boulanger PE, Neumann E, Letellier L, Conway JF. Bacteriophage T5 structure reveals similarities with HK97 and T4 suggesting evolutionary relationships. Journal of Molecular Biology. 2006;**361**:993-1002

[100] Papadopoulos S, Smith PR. The structure of the tail of the bacteriophage phi CbK. Journal of Ultrastructure Research. 1982;**80**:62-70

[101] Katsura I. Tail Assembly and Injection. Lambda II. Plainview, NY: Cold Spring Harbor; 1983. pp. 331-346

[102] Pell LG, Kanelis V, Donaldson LW, Howell PL, Davidson AR. The phage lambda major tail protein structure reveals a common evolution for long-tailed phages and the type VI bacterial secretion system. Proceedings of the National Academy of Sciences of the United States of America. 2009;**106**(11):4160-4165. DOI: 10.1073/pnas.0900044106

[103] Pell LG, Liu A, Edmonds L, Donaldson LW, Howell PL, Davidson AR. The X-ray crystal structure of the phage lambda tail terminator protein reveals the biologically relevant hexameric ring structure and demonstrates a conserved mechanism of tail termination among diverse long-tailed phages. Journal of Molecular Biology. 2009;**389**(5):938-951. DOI: 10.1016/j.jmb.2009.04.072. [Epub 2009 May 6]

[104] Chagot B, Auzat I, Gallopin M, Petitpas I, Gilquin B, Tavares P, et al. Solution structure of gp17 from the Siphoviridae bacteriophage SPP1: Insights into its role in virion assembly. Proteins. 2012;**80**(1):319-326. DOI: 10.1002/prot.23191

[105] Xiang Y, Leiman PG, Li L, Grimes S, Anderson DL, Rossmann MG. Crystallographic insights into the autocatalytic assembly mechanism of a bacteriophage tailspike. Molecular Cell. 2009;**34**:375-386. DOI: 10.1016/j.molcel.2009.04.009

[106] Xu J, Gui M, Wang D, Xiang Y. The bacteriophage φ29 tail possesses a pore-forming loop for cell membrane penetration. Nature. 2016;**534**:544-547

[107] Garcia-Doval C, Castón JR, Luque D, Granell M, Otero JM, Llamas-Saiz AL, et al. Conformational changes leading to T7 DNA delivery upon interaction with the bacterial receptor. Viruses. 2015;7(12):6424-6440. DOI: 10.3390/v7122946

[108] Steinbacher S, Baxa U, Miller S, Weintraub A, Seckler R, Huber R. Crystal structure of phage P22 tailspike protein complexed with Salmonella sp. O-antigen receptors. Proceedings of the National Academy of Sciences of the United States of America. 1996;**93**(20):10584-10588

[109] Steinbacher S, Miller S, Baxa U, Budisa N, Weintraub A, Seckler R, et al. Phage P22 tailspike protein: crystal structure of the head-binding domain at 2.3 Å, fully refined structure of the endorhamnosidase at 1.56 Å resolution, and the molecular basis of O-antigen recognition and cleavage. Journal of Molecular Biology. 1997;**267**:865-880

[110] Chang JT, Schmid MF, Haase-Pettingell C, Weigele PR, King JA, Chiu W. Visualizing the structural changes of bacteriophage Epsilon15 and its Salmonella host during infection. Journal of Molecular Biology. 2010;**402**(4):731-740. DOI: 10.1016/j.jmb.2010.07.058

[111] Taylor NM, Prokhorov NS, Guerrero-Ferreira RC, Shneider MM, Browning C, Goldie KN, et al. Structure of the T4 baseplate and its function in triggering sheath contraction. Nature. 2016;**533**:346-352

[112] Zheng W, Wang F, Taylor NMI, Guerrero-Ferreira RC, Leiman PG, Egelman EE. Refined Cryo-EM structure of the T4 tail tube: Exploring the lowest dose limit. Structure. 2017;**25**:1436-1441

[113] Fokine A, Zhang Z, Kanamaru S, Bowman VD, Aksyuk AA, Arisaka F, et al. The molecular architecture of the bacteriophage T4 neck. Journal of Molecular Biology. 2013;**425**:1731-1744

[114] Aksyuk AA, Leiman PG, Shneider MM, Mesyanzhinov VV, Rossmann MG. The structure of gene product 6 of bacteriophage T4, the hinge-pin of the baseplate. Structure. 2009;**17**:800-808

[115] Aksyuk AA, Kurochkina LP, Fokine A, Forouhar F, Mesyanzhinov VV, Tong L, et al. Structural conservation of the Myoviridae phage tail sheath protein fold. Structure. 2011;**19**:1885-1894

[116] Veesler D, Cambillau C. A common evolutionary origin for tailed bacteriophage functional modules and bacterial machineries. Microbiology and Molecular Biology Reviews. 2011;**75**:423-433

[117] Davidson AR, Cardarelli L, Pell LG, Radford DR, Maxwell KL. Long noncontractile tail machines of bacteriophages. Advances in Experimental Medicine and Biology. 2012;**726**:115-142. DOI: 10.1007/978-1-4614-0980-9_6

[118] Veesler D, Robin G, Lichière J, Auzat I, Tavares P, Bron P, et al. Crystal structure of bacteriophage SPP1 distal tail protein (gp19.1): A baseplate hub paradigm in Gram-positive infecting phages. The Journal of Biological Chemistry. 2010;**285**:36666-36673

[119] Xiang Y, Morais MC, Cohen DN, Bowman VD, Anderson DL, Rossmann MG. Crystal and cryoEM structural studies of a cell wall degrading enzyme in the bacteriophage phi29 tail. Proceedings of the National Academy of Sciences of the United States of America. 2008;**105**(28):9552-9557. DOI: 10.1073/pnas.0803787105

[120] Tang L, Marion WR, Cingolani G, Prevelige PE, Johnson JE. Three-dimensional structure of the bacteriophage P22 tail machine. The EMBO Journal. 2005;**24**:2087-2095

[121] Chang J, Weigele P, King J, Chiu W, Jiang W. Cryo-EM asymmetric reconstruction of bacteriophage P22 reveals organization of its DNA packaging and infecting machinery. Structure. 2006;**14**(6):1073-1082

[122] Tang L, Gilcrease EB, Casjens SR, Johnson JE. Highly discriminatory binding of capsid-cementing proteins in bacteriophage L. Structure. 2006;**14**(5):837-845

[123] Olia AS, Casjens S, Cingolani G. Structure of phage P22 cell envelope-penetrating needle. Nature Structural & Molecular Biology. 2007;**14**(12):1221-1226

[124] Steinbacher S, Seckler R, Miller S, Steipe B, Huber R, Reinemer P. Crystal structure of P22 tailspike protein: Interdigitated subunits in a thermostable trimer. Science. 1994;**265**:383-386

[125] Sciara G, Bebeacua C, Bron P, Tremblay D, Ortiz-Lombardia M, Lichière J, et al. Structure of lactococcal phage p2 baseplate and its mechanism of activation. Proceedings of the National Academy of Sciences of the United States of America. 2010;**107**:6852-6857

[126] Veesler D, Spinelli S, Mahony J, Lichière J, Blangy S, Bricogne G, et al. Structure of the phage TP901-1 1.8MDabaseplate suggests an alternative host adhesion mechanism. Proceedings of the National Academy of Sciences of the United States of America. 2012;**109**:8954-8958. DOI: 10.1073/pnas.1200966109

[127] Flayhan A, Vellieux FM, Lurz R, Maury O, Contreras-Martel C, Girard E, et al. Crystal structure of pb9, the distal tail protein of bacteriophage T5: A conserved structural motif among all siphophages. Journal of Virology. 2014;**88**:820-828

[128] Zivanovic Y, Confalonieri F, Ponchon L, Lurz R, Chami M, Flayhan A, et al. Insights into bacteriophage T5 structure from analysis of its morphogenesis genes and protein components. Journal of Virology. 2014;**88**:1162-1174

[129] Mitraki A, Papanikolopoulou K, van Raaij MJ. Natural triple β-stranded fibrous folds. Advances in Protein Chemistry. 2006;**73**:97-124

[130] Parent KN, Gilcrease EB, Casjens SR, Baker TS. Structural evolution of the P22-like phages: comparison of Sf6 and P22 procapsid and virion architectures. Virology. 2012;**427**:177-188

[131] Barbirz S, Muller JJ, Uetrecht C, Clark AJ, Heinemann U, Seckler R. Crystal structure of Escherichia coli phage HK620 tailspike: Podoviral tailspike endoglycosidase modules are evolutionarily related. Molecular Microbiology. 2008;**69**:303-316

[132] Rajagopala SV, Casjens S, Uetz P. The protein interaction map of bacteriophage lambda. BMC Microbiology. 2011;**11**:213

[133] Morais MC, Tao Y, Olson NH, Grimes S, Jardine PJ, Anderson DL, et al. Cryoelectron-microscopy image reconstruction of symmetry mismatches in bacteriophage φ29. Journal of Structural Biology. 2001;**135**:38-46

[134] Farley MM, Tu J, Kearns DB, Molineux IJ, Liu J. Ultrastructural analysis of bacteriophage Φ29 during infection of Bacillus subtilis. Journal of Structural Biology. 2017;**197**(2):163-171. DOI: 10.1016/j.jsb.2016.07.019

[135] Tao Y, Olson NH, Xu W, Anderson DL, Rossmann MG, Baker TS. Assembly of a tailed bacterial virus and its genome release studied in three dimensions. Cell. 1998;**95**:431-437

[136] Leiman PG, Kanamaru S, Mesyanzhinov VV, Arisaka F, Rossmann MG. Structure and morphogenesis of bacteriophage T4. Cellular and Molecular Life Sciences. 2003;**60**(11):2356-2370

[137] Leiman PG, Arisaka F, van Raaij MJ, Kostyuchenko VA, Aksyuk AA, Kanamaru S, et al. Morphogenesis

of the T4 tail and tail fibers. Virology Journal. 2010;7:355. DOI: 10.1186/1743-422X-7-355

[138] Arisaka F, Yap ML, Kanamaru S, Rossmann MG. Molecular assembly and structure of the bacteriophage T4 tail. Biophysical Reviews. 2016;**8**(4):385-396. DOI: 10.1007/s12551-016-0230-x

[139] Crawford JT, Goldberg EB. The function of tail fibers in triggering baseplate expansion of bacteriophage T4. Journal of Molecular Biology. 1980;**139**(4):679-690

[140] Kostyuchenko VA, Leiman PG, Chipman PR, Kanamaru S, van Raaij MJ, Arisaka F, et al. Three-dimensional structure of bacteriophage T4 baseplate. Nature Structural Biology. 2003;**10**(9):688-693

[141] Yap ML, Rossmann MG. Structure and function of bacteriophage T4. Future Microbiology. 2014;**9**(12): 1319-1327. DOI: 10.2217/fmb.14.91

[142] Yap ML, Klose T, Arisaka F, Speirc JA, Veeslerc D, Fokine A, et al. The role of bacteriophage T4 baseplate in regulating assembly and infection. Proceedings of the National Academy of Sciences of the United States of America. 2016;**113**(10):2654-2659

[143] Kanamaru S, Leiman PG, Kostyuchenko VA, Chipman PR, Mesyanzhinov VV, Arisaka F, et al. Structure of the cell-puncturing device of bacteriophage T4. Nature. 2002;**415**:553-557. DOI: 10.1038/415553a

[144] Shneider MM, Buth SA, Ho BT, Basler M, Mekalanos JJ, Leiman PG. PAAR-repeat proteins sharpen and diversify the type VI secretion system spike. Nature. 2013;**500**:350-353. DOI: 10.1038/nature1245

[145] Tavares P, Zinn-Justin S, Orlova EV. Genome gating in tailed bacteriophage capsids. Advances in Experimental Medicine and Biology. 2012;**726**:585-600. DOI: 10.1007/978-1-4614-0980-9_25. [Review. PMID: 22297531]

[146] Ando H, Lemire S, Pires DP, Lu TK. Engineering modular viral scaffolds for targeted bacterial population editing. Cell Systems. 2015;**1**(3):187-196

[147] Abedon ST. Ecology of anti-biofilm agents II: Bacteriophage exploitation and biocontrol of biofilm bacteria. Pharmaceuticals (Basel). 2015;**8**(3): 559-589. DOI: 10.3390/ph8030559

Chapter 3

Biotechnology Tools Derived from the Bacteriophage/Bacteria Arms Race

Vitor B. Pinheiro

Abstract

The long association and intense competition between bacteria and their viruses have created a fertile ground for evolution to develop numerous tools for DNA modification, assembly and degradation. Many of these tools underpin the past 50 years of molecular biology, and others show great potential in shaping the next 50 years of the field. Here, I present some of the tools that have come out of the bacteria-bacteriophage arms race and discuss some of the concepts that may shape their future use. Molecular biology remains a fast-growing area increasingly limited solely by researcher ingenuity.

Keywords: molecular biology tools, orthogonality, DNA modifications

1. Introduction

The relationship between bacteriophages and bacteria is often explained in terms of an arms race: each 'developing' measures and countermeasures for attacking and defending itself from the other [1, 2]. The imagery of an arms race is a powerful metaphor to summarise the relationship between possibility and availability that have constrained the emergence, evolution and diversification of life on Earth.

Life is limited by what is possible. For instance, life can only exist because chemical information storage is possible. On the other hand, life as we know it has evolved around DNA and RNA because they are informational molecules that could function in the environment of the early Earth and whose building blocks were readily available.

The relationship between bacteria and bacteriophages is similarly constrained. The emergence (or availability) of bacterial cells capable of establishing a rich internal environment (compared to the outside of the cell) creates the possibility for other organisms to evolve predatory or parasitic survival strategies, including bacteriophages. Once phages emerge, they alter the dynamics of the ecological niche and create an advantage to bacterial hosts that can reduce the success of phage infection—whether by hindering phage access to the cell cytoplasm, by interfering with phage survival or replication in the cell or by interfering with phage maturation and release [3].

Bacterial defences arise from any function already available in the host (e.g. uracil-DNA glycosylase involved in DNA repair) or that can be co-opted from available genetic resources in the cell or in the environment

IntechOpen

(e.g. restriction-modification systems). Defences can also emerge from loss of function (e.g. mutations to the maltose porin LamB in *E. coli*, which make it resistant to bacteriophage λ infections [4]).

Given the prevalence of bacteria and phages in the environment, and given the evolutionary scale time of their arms race, the variety, complexity and efficiency of these attack and defence strategies are huge and can range from silent integration into the host genome (i.e. lysogeny) to enacting a hostile molecular takeover of the bacterial host cell. Despite our current efforts to map the genetic diversity available on Earth, it remains likely that new strategies are still to be identified and characterised.

Nonetheless, many of these defence and attack strategies have also been harnessed for biotechnology applications, significantly beyond the simple use of bacteriophages (or bacteriophage proteins) as bacterial control agents [5–7]. Restriction-modification (RM) systems found in bacteria were among the very first tools isolated from the bacteria-phage arms race [8, 9]. They *de facto* represent the start of modern molecular biology, and they have remained key tools for over 50 years (**Figure 1**).

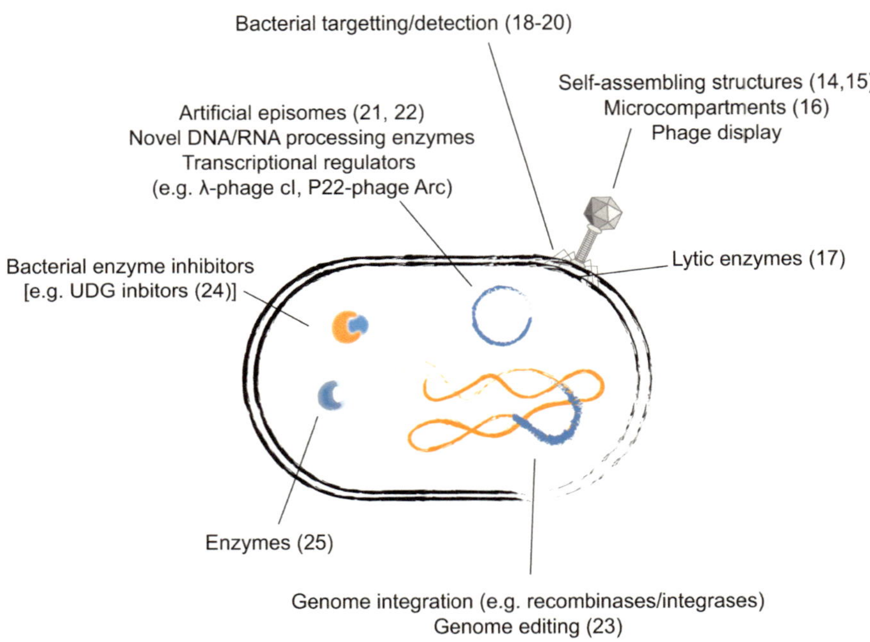

Figure 1.
Potential biotechnology applications derived from bacteriophages. Almost every aspect of the bacteriophage life cycle can be exploited for the development of biotechnology tools. Aside from their natural bactericidal role, phages can be used for the delivery of engineered circuits [10, 11] and in directed evolution through phage display [12] and PACE [13]. Individual phage systems have been also successfully developed as tools.

More recently, another tool derived from bacterial defence has been harnessed, with potentially equally transformative impact on how we interact with biology: clustered regularly interspersed palindromic repeats (CRISPR) [26–28]. CRISPR forms part of an adaptive immunity system in prokaryotes, but it is being harnessed to deliver a wide range of research and therapeutic tools.

Although RM systems and CRISPR are deservedly acknowledged as having a significant impact on molecular biology and biotechnology, many other tools have been or are being developed based on components isolated from the bacteria-phage arms race. This chapter focuses on some of those tools—their mechanisms, current and potential applications.

2. Common molecular biology tools and orthogonality

Because of the wide range of bacteriophage infection strategies available, it becomes difficult to introduce simple classification without recreating the complexity of approaches taken by phages. For instance, bacteriophage promoters can rely exclusively on host proteins (e.g. T4 early promoters), on a mixture of host and phage proteins (e.g. T4 middle promoters or P_L promoter from λ phage) or exclusively on phage-derived factors (e.g. T7 RNA polymerase promoters). This provides a continuum that can be further dissected by analysing the mechanism of the hybrid promoters, with their specific host and phage dependencies.

That continuum maps how independent a phage system is from the host while still active within the host, i.e. it is a measure of the orthogonality of the system. Having evolved to survive in a changing environment, bacteria have complex layers of gene expression regulation with multiple feedback systems which are not necessarily easy to control independently, despite our advances in understanding bacterial metabolism [29, 30]. In that context, phage systems that have reduced dependencies on the host machinery (i.e. increased orthogonality) provide isolated systems that can be simpler to regulate and are, at least in part, shielded from variations in the cellular machinery—an approach that has dominated biotechnology until recently.

Many of the common phage-derived biotechnology tools have been developed from such systems, none more so than T7 RNA polymerase [31]. Isolated from T7 bacteriophage, this monomeric RNA polymerase can recognise a specific promoter sequence. The core T7 promoter sequence (TAATACGACTCACTATAG) is sufficient to trigger transcription in cells harbouring a T7 RNA polymerase gene. This is the strategy set up behind pET vectors, which contain a T7 promoter and rely on an *E. coli* host carrying a T7 RNA polymerase under an inducible promoter (e.g. *lacUV*)—usually the result of the introduction of a DE3 phage [32, 33].

Nevertheless, the context of the T7 promoter can have a significant impact on the expression level of the downstream genes, and at high polymerase concentrations, it is possible to drive transcription from suboptimal promoters—highlighting that the orthogonality of the system is limited.

Given its monomeric structure and orthogonal role in cellular transcription, T7 RNA polymerase became not only a useful tool in biotechnology but also an important model system for the study of transcription (reviewed in [34]). Because of its orthogonality, T7 RNA polymerase (and its promoter) can be harnessed for the regulation of transcription in a wide range of hosts beyond *E. coli*, including Gram-positive bacterial hosts [35], yeast [36, 37] and human cells [38, 39].

Its role in the regulation of transcription has also been expanded through the creation of more complex systems using split T7 RNA polymerase proteins. Surprisingly for a mesophilic highly dynamic enzyme, T7 RNA polymerase can be expressed in two [40, 41] or more [42] fragments that *in vivo* are able to reassemble and function as viable RNA polymerases.

While orthogonality can be a desirable feature for *in vivo* applications, it is wholly unnecessary for in vitro applications, where the key constraint lies on identifying reaction conditions in which expressed and purified proteins are sufficiently active to carry out the desired function. That is the case with T4 DNA ligase and T4 polynucleotide kinase, which were originally isolated from the *E. coli* T4 phage and remain important tools in molecular biology.

T4 DNA ligase has a central role in the replication and repair of the phage genome during its infection of *E. coli* [43]. This also entails coping with DNA modifications such as the full substitution of cytosine for 5-hydroxymethylcytosine and glucosyl-5-hydroxymethylcytosine that naturally occurs *in vivo* [44, 45]—this is a phage defence mechanism discussed below. Although its structure has only recently been determined [46], the mechanism of action of this enzyme has long been characterised [47]. Even in the absence of other phage genes, it is active *in vivo* [48], but its main application in molecular biology remains its in vitro activity that, coupled with restriction endonucleases, has underpinned modern molecular biology—allowing a molecular cut and paste approach to DNA assembly.

Ligase in vitro activity, particularly its ability to accept modified ligands, has been extensively explored for the assembly of heavily modified DNA sequences for aptamer selection [49, 50] and to explore a wider range of nucleic acid modifications, such as sugar-modified nucleic acids [51].

3. Second-generation tools and applications

The combination of different enzymatic functions has created novel applications, such as the large-scale DNA assembly in Gibson assembly through the combination of exonuclease, polymerase and ligase activities [52]. However, a whole range of novel applications are possible by harnessing additional bacteriophage proteins that have not yet been extensively explored.

Recombinases and integrases, enzymes that catalyse the sequence-specific insertion of a phage genome into the host chromosome [53, 54], were identified early in bacteriophage research (e.g. the λ integrase) but were not immediately harnessed for applications. These came substantially later once recombinases were shown to facilitate DNA assembly, whether by increasing the efficiency of subcloning such as in Gateway cloning [55], or enabling multipart assemblies needed for metabolic engineering [56].

In general, recombinases bind DNA specifically, as dimers, on recognition sites that are relatively short (usually between 30 and 50 bases) and partially palindromic, termed *attP* and *attB* (originally to distinguish phage and bacterial origins). These two sites have different sequences which contribute to making the recombination process unidirectional. The recombinases facilitate the break and religation of double-stranded DNA within the *att* sites resulting in chimeric sites that are then labelled *attR* and *attL* (from right and left sides, respectively). Insertion of a phage genome into the host is efficient and stable, but it can also be reversed with the contribution of a single host factor (reviewed in [57]).

Because of the high efficiency of integration, recombinases have been also developed as systems to facilitate homologous recombination in higher eukaryotes, such as mammalian cells [58], *Drosophila* organisms [59] and plants (reviewed in [60]). In the latter, recombinases were of particular interest because of their potential to remove transformation and selection markers from engineered crops, leaving behind only the genes responsible for the engineered trait. This idea of using recombinases to directly alter an organism's genome has been vastly expanded in the Yeast 2.0 project, through the design implementation of multiple *loxP* (the equivalent to *att* sites for Cre recombinase) in the synthetic yeast genome that can be activated, leading to large-scale genome rearrangement—termed synthetic chromosome rearrangement and modification by loxP-mediated evolution (SCRaMbLE) [61].

By enabling controllable chromosome rearrangements between the designed *loxP* sites, a synthetic yeast can delete, duplicate and reorder many of its genes, allowing *in vivo* selection for desirable traits such as increased alkali tolerance [62]. Alternatively, the system can be coupled to heterologously expressed genes allowing the rapid optimisation of pathways [63].

Like SCRaMbLE, protein-directed evolution relies on cycles of mutagenesis (to introduce diversity into a population) and selection (to reduce diversity towards functional proteins) [64] which in some platforms can be achieved continuously *in vivo*—e.g. in PACE [13] or in some continuous culture approaches [65]. In both examples, mutation can be controlled by stressors or the induction of error-prone replication but are not necessarily limited to the area of interest (e.g. a single gene). Greater control of targeting is possible and has been reported through the use of an error-prone PolI [66]—which is necessary for the replication of some bacterial plasmids and can be used to drive diversification *in vivo* in the vicinity of plasmid replication initiation—and protein fusions that target an error-prone polymerase to a particular region of the genome [e.g. EvolvR and MutaT7 (reviewed in [67])].

However, one such system, termed diversity-generating retroelements (DGRs), has naturally emerged in bacteriophages and was first implicated in the tropism switching of *Bordetella* phages [68, 69] but that has since been identified in a wide range of bacterial and archaeal genomes [70, 71]. The system relies on an error-prone reverse transcriptase (RT) and on two DNA repeats—one operating as a template (template repeat) and the other as the target (variable repeat). RNA synthesised from the template is used to guide the error-prone synthesis of a DNA by the RT, which also coordinates its insertion at the target site. The error-profile of DGR RTs results in adenines being replaced with other bases, creating a directionality to the evolution that is always anchored by the template repeat [72]. Nevertheless, changes in DNA sequences involved in the targeting of the retroelement, termed initiation of mutagenic homing (IMH), can lead to both template and variable repeats being allowed to evolve at different rates—removing some of the directionality in evolution and freeing the system to explore the sequence space more thoroughly [68].

Despite its potential, it remains to be seen whether such a system can be harnessed for protein engineering. If the DGR systems can themselves be engineered, their targeting and error rate may be amenable to modulation opening possibilities to compete with the most recent Cas9-derived gene editors [73].

4. Xenobiotic nucleic acids

Chemical modification of the phage's own genome is a widespread strategy that emerged multiple times in evolution to circumvent (or at least slow down) bacterial defence mechanisms that target the invading DNA for degradation: restriction endonucleases, exonucleases and CRISPR-Cas systems [44, 74]. Those modifications have been reported not only on the nucleobases, akin to eukaryotic epigenetic markers, but also on the nucleic acid backbone.

Reported nucleobase modifications suggest that the overwhelming majority of any such modifications is targeted to the C5 position in pyrimidines. They range from small modifications, such as methylation, to bulky modifications, such as putrescine and even carbohydrate moieties. While such modifications have long been known, new sequencing platforms capable of reading the DNA sequence without an amplification step, such as nanopore sequencing, hold great promise in enabling a more systematic mapping and characterisation of DNA modifications in phage genomes (**Figure 2**) [75].

DNA modifications, particularly modifications that bring chemical functionality not available in natural bases, such as glycosylation in *Bacillus subtilis* SP-15 phage [76], can be harnessed for function as has been achieved through the chemical modification of DNA bases and systematic evolution of ligands by exponential enrichment (SELEX) [50, 77]. Despite characterisation of the biosynthetic pathway for multiple-phage DNA modification systems, none have been implemented in vitro for applications.

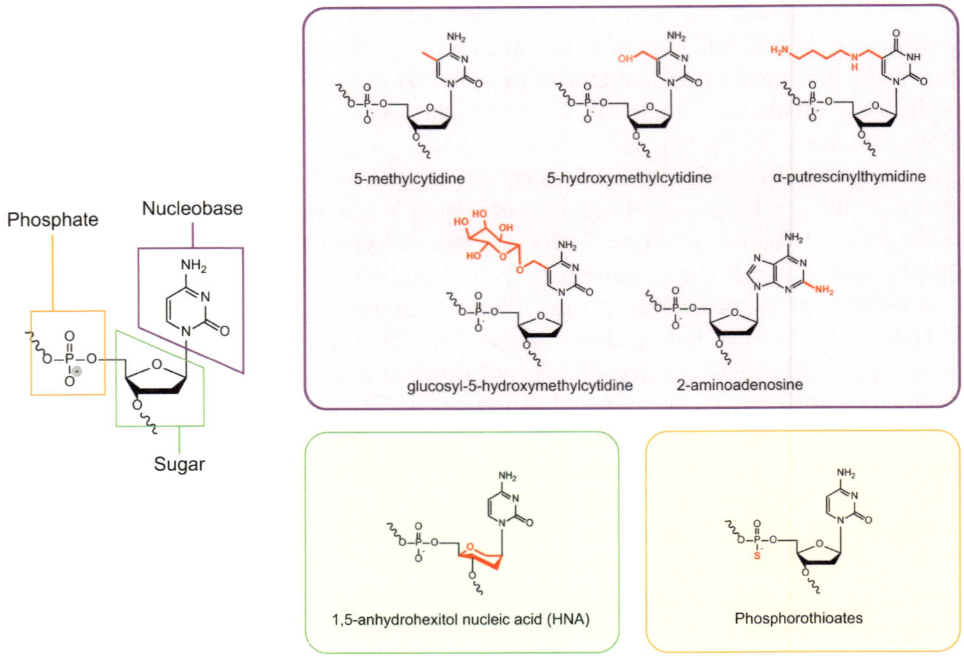

Figure 2.
Examples of XNAs in phages. Nucleobases are composed of three different chemical moieties, and change to any one of them has the potential to alter duplex conformation and how easily they are recognised by natural nucleic acid processing enzymes [14–25]. The base, sugar and backbone discussed in the chapter are shown here.

One potential bottleneck lies on how the phage and the host handle those chemically modified DNAs. Upon infection, the mature phage DNA needs to be modified if that is an evolutionary strategy being exploited to slow down or avoid *in vivo* degradation. On the other hand, those chemical modifications can affect DNA structure and biophysical properties which may also be detrimental to bacteriophage replication—since this would require a DNA polymerase capable of processing such heavily modified genomes.

It is known that at least in some cases, this chemical cloaking is removed upon infection, such as in SP-15 [78], before unmodified DNA is replicated *in vivo*. But, given viral polymerases more permissive substrate specificity, it is possible that some systems can be replicated directly by highly adapted phage polymerases—either to DNA followed by reinstalling the chemical modifications or directly from modified DNA to modified DNA. In the case of SP-15, the bulkier modifications are rapidly removed prior to replication of DNA [78]. T4 seems to follow a similar pattern where glycosylation is removed, and DNA is replicated containing only the simpler 5-hydroxymethylation modification. This is further supported by the biochemical evidence that glycosylation is 'installed' on the replicated T4 DNA [44, 74]. Nevertheless, early T4 transcription occurs rapidly, and it is carried out by the host RNA polymerase, suggesting that natural RNA polymerases can still use the hypermodified bases in that template.

Notably, although phage polymerases replicate phage DNA *in vivo* harbouring simple chemical modifications, such as (in *Synechocystis* S-2 L phage) 2-aminoadenine [79] and uracil (in *Bacillus* phages AR9, PBS1 and PBS2, *Yersinia* phage PhiR1-37 and *Staphylococcus* phage S6), no viral polymerase has been described that is capable of selectively incorporating the modified bases. That is, although some bacteriophages make use of modified nucleobases and have evolved systems that lead to 100% incorporation of the modified bases in their genomes, their DNA polymerases have not specialised towards being able to only incorporate the modified nucleobases—they remain able to recognise unmodified triphosphates.

Still, the increased substrate flexibility of phage DNA polymerases may at least in part justify why a *Bacillus subtilis* Phi29 DNA polymerase required a single mutation for the synthesis of anhydrohexitol nucleic acids (HNA) [80] while an archaeal enzyme required in excess of seven mutations [81].

5. Conclusion

Bacteriophages remain a rich source of novel functionalities that can be harnessed to advance molecular biology (and synthetic biology). The examples here provided represent only a small fraction of the potential applications available, which also include medical applications from phage proteins [82, 83] and engineered phages [10, 11, 84].

In addition, bacteriophages have had a close relationship with directed evolution, either as a vehicle such as in phage display [85, 86] or by providing (in addition to the examples above) proteins to accelerate strain engineering, such as in multiplex automated genome engineering (MAGE) [87].

Finally, bacteriophages may also become an important tool in harnessing new non-model organisms in synthetic biology, as pre-optimised DNA delivery nanomachines for custom circuits.

Author details

Vitor B. Pinheiro[1,2]

1 Rega Institute for Medical Research, KU Leuven, Leuven, Belgium

2 Department of Structural and Molecular Biology, University College London, London, UK

*Address all correspondence to: vitor.pinheiro@kuleuven.be

References

[1] Koonin EV, Makarova KS, Wolf YI. Evolutionary genomics of defense systems in Archaea and Bacteria. Annual Review of Microbiology. 2017;**71**(1):233-261. DOI: 10.1146/annurev-micro-090816-093830

[2] Labrie SJ, Samson JE, Moineau S. Bacteriophage resistance mechanisms. Nature Reviews. Microbiology. 2010;**8**:317. DOI: 10.1038/nrmicro2315

[3] Seed KD. Battling phages: How bacteria defend against viral attack. PLoS Pathogens. 2015;**11**(6):e1004847. DOI: 10.1371/journal.ppat.1004847

[4] Chatterjee S, Rothenberg E. Interaction of bacteriophage l with Its *E. coli* receptor, LamB. Viruses. 2012;**4**:3162-3178

[5] Lemire S, Yehl KM, Lu TK. Phage-based applications in synthetic biology. Annual Review of Virology. 2018;**5**(1):453-476. DOI: 10.1146/annurev-virology-092917-043544

[6] Salmond GPC, Fineran PC. A century of the phage: Past, present and future. Nature Reviews. Microbiology. 2015;**13**:777. DOI: 10.1038/nrmicro3564

[7] Kortright KE, Chan BK, Koff JL, Turner PE. Phage therapy: A renewed approach to combat antibiotic-resistant bacteria. Cell Host & Microbe. 2019;**25**(2):219-232. Available from: https://www.sciencedirect.com/science/article/pii/S1931312819300526?via%3Dihub

[8] Tock MR, Dryden DTF. The biology of restriction and anti-restriction. Current Opinion in Microbiology. 2005;**8**(4):466-472

[9] Pingoud A, Wilson GG, Wende W. Type II restriction endonucleases—A historical perspective and more. Nucleic Acids Research. 2014;**42**(12):7489-7527. DOI: 10.1093/nar/gku447

[10] Ando H, Lemire S, Pires DP, Lu TK. Engineering modular viral scaffolds for targeted bacterial population editing. Cell Systems. 2015;**1**(3):187-196

[11] Bikard D, Euler CW, Jiang W, Nussenzweig PM, Goldberg GW, Duportet X, et al. Exploiting CRISPR-cas nucleases to produce sequence-specific antimicrobials. Nature Biotechnology. 2014;**32**(11):1146-1150

[12] Fernandez-Gacio A, Uguen M, Fastrez J. Phage display as a tool for the directed evolution of enzymes. Trends in Biotechnology. 2003;**21**(9):408-414

[13] Esvelt KM, Carlson JC, Liu DR. A system for the continuous directed evolution of biomolecules. Nature. 2011;**472**(7344):499. DOI: 10.1038/nature09929

[14] Moon J-S, Kim W-G, Kim C, Park G-T, Heo J, Yoo SY, et al. M13 bacteriophage-based self-assembly structures and their functional capabilities. Mini-Reviews in Organic Chemistry. 2015;**12**(3):271-281

[15] Han J, Devaraj V, Kim C, Kim W-G, Han D-W, Hong SW, et al. Fabrication of self-assembled nanoporous structures from a self-templating M13 bacteriophage. ACS Applied Nano Materials. 2018;**1**(6):2851-2857. DOI: 10.1021/acsanm.8b00500

[16] Krupovic M, Koonin EV. Multiple origins of viral capsid proteins from cellular ancestors. Proceedings of the National Academy of Sciences of the United States of America. 2017;**114**(12):E2401-E2410

[17] Wang I-N, Smith DL, Young R. Holins: The protein clocks of bacteriophage infections. Annual Review of Microbiology. 2000;**54**(1):799-825

[18] Denyes JM, Dunne M, Steiner S, Mittelviefhaus M, Weiss A, Schmidt H, et al. Modified bacteriophage S16 long tail fiber proteins for rapid and specific immobilization and detection of *Salmonella* cells. Applied and Environmental Microbiology. 2017;**83**(12):e00277-17

[19] Schmelcher M, Loessner MJ. Bacteriophage: Powerful tools for the detection of bacterial pathogens. In: Principles of Bacterial Detection: Biosensors, Recognition Receptors and Microsystems. New York: Springer; 2008. p. 731-754. ISBN: 978-0-387-75112-2

[20] Dunne M, Loessner MJ. Modified bacteriophage tail fiber proteins for labeling, immobilization, capture, and detection of bacteria. Methods in Molecular Biology. 2019;**1918**:67-86

[21] Ravikumar A, Arrieta A, Liu CC. An orthogonal DNA replication system in yeast. Nature Chemical Biology. 2014;**10**(3):175-177

[22] Mencía M, Gella P, Camacho A, de Vega M, Salas M. Terminal protein-primed amplification of heterologous DNA with a minimal replication system based on phage Phi29. Proceedings of the National Academy of Sciences of the United States of America. 2011;**108**(46):18655-18660. Available from: http://www.pubmedcentral.nih.gov/articlerender.fcgi?artid=3219123&tool=pmcentrez&rendertype=abstract

[23] Carr PA, Wang HH, Sterling B, Isaacs FJ, Lajoie MJ, Xu G, et al. Enhanced multiplex genome engineering through co-operative oligonucleotide co-selection. Nucleic Acids Research. 2012;**40**(17):e132

[24] Serrano-Heras G, Ruiz-Masó JA, Del Solar G, Espinosa M, Bravo A, Salas M. Protein p56 from the *Bacillus subtilis* phage Φ29 inhibits DNA-binding ability of uracil-DNA glycosylase. Nucleic Acids Research. 2007;**34**(16):5393-5401

[25] Thompson LR, Zeng Q, Kelly L, Huang KH, Singer AU, Stubbe JA, et al. Phage auxiliary metabolic genes and the redirection of cyanobacterial host carbon metabolism. Proceedings of the National Academy of Sciences of the United States of America. 2011;**108**(39):E757-E764

[26] Lander ES. The heroes of CRISPR. Cell. 2016;**164**(1-2):18-28. Available from: https://www.sciencedirect.com/science/article/pii/S0092867415017055?via%3Dihub#bib34

[27] Mojica FJM, Juez G, Rodriguez-Valera F. Transcription at different salinities of *Haloferax mediterranei* sequences adjacent to partially modified PstI sites. Molecular Microbiology. 1993;**9**(3):613-621. DOI: 10.1111/j.1365-2958.1993.tb01721.x

[28] Jansen R, Embden JDA van, Gaastra W, Schouls LM. Identification of genes that are associated with DNA repeats in prokaryotes. Molecular Microbiology. 2002;**43**(6):1565-1575. DOI: 10.1046/j.1365-2958.2002.02839.x

[29] Gibson DG, Benders GA, Andrews-Pfannkoch C, Denisova EA, Baden-Tillson H, Zaveri J, et al. Complete chemical synthesis, assembly, and cloning of a mycoplasma genitalium genome. Science. 2008;**319**(5867):1215-1220

[30] Orth JD, Conrad TM, Na J, Lerman JA, Nam H, Feist AM, et al. A comprehensive genome-scale reconstruction of *Escherichia coli* metabolism-2011. Molecular Systems Biology. 2011;**7**(1):535

[31] Tabor S, Richardson CC. A bacteriophage T7 RNA polymerase/promoter system for controlled exclusive expression of specific genes.

Proceedings of the National Academy of Sciences of the United States of America. 1985;**82**(4):1074-1078

[32] Rosenberg AH, Lade BN, Dao-shan C, Lin SW, Dunn JJ, Studier FW. Vectors for selective expression of cloned DNAs by T7 RNA polymerase. Gene. 1987;**56**(1):125-135

[33] Giordano TJ, Deuschle U, Bujard H, McAllister WT. Regulation of coliphage T3 and T7 RNA polymerases by the lac represser-operator system. Gene. 1989;**84**(2):209-219

[34] Sousa R. T7 RNA polymerase. In: Encyclopedia of Biological Chemistry. 2nd ed. Elsevier; 2013. p. 355-359. ISBN: 9780123786319

[35] Kortmann M, Kuhl V, Klaffl S, Bott M. A chromosomally encoded T7 RNA polymerase-dependent gene expression system for *Corynebacterium glutamicum*: Construction and comparative evaluation at the single-cell level. Microbial Biotechnology. 2015;**8**(2):253-265

[36] Hobl B, Hock B, Schneck S, Fischer R, Mack M. Bacteriophage T7 RNA polymerase-based expression in *Pichia pastoris*. Protein Expression and Purification. 2013;**92**(1):100-104

[37] Benton BM, Eng WK, Dunn JJ, Studier FW, Sternglanz R, Fisher PA. Signal-mediated import of bacteriophage T7 RNA polymerase into the *Saccharomyces cerevisiae* nucleus and specific transcription of target genes. Molecular and Cellular Biology. 1990;**10**(1):353-360

[38] Aoki Y, Aizaki H, Shimoike T, Tani H, Ishii K, Saito I, et al. A human liver cell line exhibits efficient translation of HCV RNAs produced by a recombinant adenovirus expressing T7 RNA polymerase. Virology. 1998;**250**(1):140-150

[39] Dunn JJ, Krippl B, Bernstein KE, Westphal H, William SF. Targeting bacteriophage T7 RNA polymerase to the mammalian cell nucleus. Gene. 1988;**68**(2):259-266

[40] Schaerli Y, Gili M, Isalan M. A split intein T7 RNA polymerase for transcriptional AND-logic. Nucleic Acids Research. 2014;**42**(19):12322-12328

[41] Shis DL, Bennett MR. Library of synthetic transcriptional AND gates built with split T7 RNA polymerase mutants. Proceedings of the National Academy of Sciences of the United States of America. 2013;**110**(13):5028-5033

[42] Segall-Shapiro TH, Meyer AJ, Ellington AD, Sontag ED, Voigt CA. A 'resource allocator' for transcription based on a highly fragmented T7 RNA polymerase. Molecular Systems Biology. 2014;**10**(7):742

[43] Mueser TC, Hinerman JM, Devos JM, Boyer RA, Williams KJ. Structural analysis of bacteriophage T4 DNA replication: A review in the virology journal series on bacteriophage T4 and its relatives. Virology Journal. 2010;7(1):359

[44] Weigele P, Raleigh EA. Biosynthesis and function of modified bases in bacteria and their viruses. Chemical Reviews. 2016;**116**(20):12655-12687

[45] Vlot M, Houkes J, SJA L, Swarts DC, Zheng P, Kunne T, et al. Bacteriophage DNA glucosylation impairs target DNA binding by type I and II but not by type V CRISPR–Cas effector complexes. Nucleic Acids Research. 2018;**46**(2):873-885

[46] Shi K, Bohl TE, Park J, Zasada A, Malik S, Banerjee S, et al. T4 DNA ligase structure reveals a prototypical ATP-dependent ligase with a unique mode of sliding clamp

interaction. Nucleic Acids Research. 2018;**46**(19):10474-10488

[47] Rossi R, Montecucco A, Ciarrocchi G, Biamonti G. Functional characterization of the T4 DNA ligase: A new insight into the mechanism of action. Nucleic Acids Research. 1997;**25**(11):2106-2113

[48] Su T, Liu F, Chang Y, Guo Q, Wang J, Wang Q, et al. The phage T4 DNA ligase mediates bacterial chromosome DSBs repair as single component non-homologous end joining. Synthetic and Systems Biotechnology. 2019;**4**(2):107-112

[49] Hili R, Niu J, Liu DR. DNA ligase-mediated translation of DNA into densely functionalized nucleic acid polymers. Journal of the American Chemical Society. 2013;**135**(1):98-101. DOI: 10.1021/ja311331m

[50] Kong D, Yeung W, Hili R. In vitro selection of diversely functionalized aptamers. Journal of the American Chemical Society. 2017;**139**(40):13977-13980

[51] Kestemont D, Renders M, Leonczak P, Abramov M, Schepers G, Pinheiro VB, et al. XNA ligation using T4 DNA ligase in crowding conditions. Chemical Communications. 2018;**54**(49):6408-6411. DOI: 10.1039/C8CC02414F

[52] Gibson DG, Young L, Chuang R-Y, Venter JC, Hutchison CA, Smith HO. Enzymatic assembly of DNA molecules up to several hundred kilobases. Nature Methods. 2009;**6**(5):343-345

[53] Landy A. The λ integrase site-specific recombination pathway. In: Mobile DNA III. American Society of Microbiology; 2015. p. 91-118. ISSN: 2165-0497

[54] Groth AC, Calos MP. Phage integrases: Biology and applications. Journal of Molecular Biology. 2004;**335**(3):667-678

[55] Hartley JL, Temple GF, Brasch MA. DNA cloning using *in vitro* site-specific recombination. Genome Research. 2000;**10**(11):1788-1795

[56] Colloms SD, Merrick CA, Olorunniji FJ, Stark WM, Smith MCM, Osbourn A, et al. Rapid metabolic pathway assembly and modification using serine integrase site-specific recombination. Nucleic Acids Research. 2014;**42**(4):e23

[57] Rutherford K, Van Duyne GD. The ins and outs of serine integrase site-specific recombination. Current Opinion in Structural Biology. 2014;**24**:125-131

[58] Thyagarajan B, Olivares EC, Hollis RP, Ginsburg DS, Calos MP. Site-specific genomic integration in mammalian cells mediated by phage φC31 integrase. Molecular and Cellular Biology. 2001;**21**(12):3926-3934

[59] Groth AC, Fish M, Nusse R, Calos MP. Construction of transgenic *Drosophila* by using the site-specific integrase from phage φC31. Genetics. 2004;**166**(4):1775-1782

[60] Ow DW. The long road to recombinase-mediated plant transformation. Plant Biotechnology Journal. 2016;**14**(2):441-447

[61] Shen Y, Stracquadanio G, Wang Y, Yang K, Mitchell LA, Xue Y, et al. SCRaMbLE generates designed combinatorial stochastic diversity in synthetic chromosomes. Genome Research. 2016;**26**(1):36-49

[62] Ma L, Li Y, Chen X, Ding M, Wu Y, Yuan YJ. SCRaMbLE generates evolved yeasts with increased alkali tolerance. Microbial Cell Factories. 2019;**18**(1):52

[63] Liu W, Pham NT, Tuck L, Pérez-Pi I, Shen Y, French C, et al. Rapid pathway

prototyping and engineering using *in vivo* and *in vivo* synthetic genome SCRaMbLE-in methods. Nature Communications. 2018;**9**(1):1936

[64] Tizei PAG, Csibra E, Torres L, Pinheiro VB. Selection platforms for directed evolution in synthetic biology. Biochemical Society Transactions. 2016;**44**(4):1165-1175. DOI: 10.1042/BST20160076

[65] Marlière P, Patrouix J, Döring V, Herdewijn P, Tricot S, Cruveiller S, et al. Chemical evolution of a bacterium's genome. Angewandte Chemie, International Edition. 2011;**50**(31):7109-7114

[66] Camps M, Naukkarinen J, Johnson BP, Loeb LA. Targeted gene evolution in *Escherichia coli* using a highly error-prone DNA polymerase I. Proceedings of the National Academy of Sciences of the United States of America. 2003;**100**(17):9727-9732

[67] Yang J, Kim B, Kim GY, Jung GY, Seo SW. Synthetic biology for evolutionary engineering: From perturbation of genotype to acquisition of desired phenotype. Biotechnology for Biofuels. 2019;**12**(1):113

[68] Doulatov S, Hodes A, Dal L, Mandhana N, Liu M, Deora R, et al. Tropism switching in *Bordetella* bacteriophage defines a family of diversity-generating retroelements. Nature. 2004;**431**(7007):476-481

[69] Liu M, Deora R, Doulatov SR, Gingery M, Eiserling FA, Preston A, et al. Reverse transcriptase-mediated tropism switching in *Bordetella* bacteriophage. Science. 2002;**295**(5562):2091-2094

[70] Paul BG, Burstein D, Castelle CJ, Handa S, Arambula D, Czornyj E, et al. Retroelement-guided protein diversification abounds in vast lineages of bacteria and archaea. Nature Microbiology. 2017;**2**(6):17045

[71] Paul BG, Bagby SC, Czornyj E, Arambula D, Handa S, Sczyrba A, et al. Targeted diversity generation by intraterrestrial archaea and archaeal viruses. Nature Communications. 2015;**6**(1):6585

[72] Guo H, Tse LV, Barbalat R, Sivaamnuaiphorn S, Xu M, Doulatov S, et al. Diversity-generating retroelement homing regenerates target sequences for repeated rounds of codon rewriting and protein diversification. Molecular Cell. 2008;**31**(6):813-823

[73] Anzalone AV, Randolph PB, Davis JR, Sousa AA, Koblan LW, Levy JM, et al. Search-and-replace genome editing without double-strand breaks or donor DNA. Nature. 2019;**576**(7785):149-157. DOI: 10.1038/s41586-019-1711-4

[74] Warren RAJ. Modified bases in bacteriophage DNAs. Annual Review of Microbiology. 1980;**34**(1):137-158

[75] Liu Q, Fang L, Yu G, Wang D, Xiao C-L, Wang K. Detection of DNA base modifications by deep recurrent neural network on Oxford nanopore sequencing data. Nature Communications. 2019;**10**(1):2449. DOI: 10.1038/s41467-019-10168-2

[76] Ehrlich M, Ehrlich KC. A novel, highly modified, bacteriophage DNA in which thymine is partly replaced by a phosphoglucuronate moiety covalently bound to 5-(4′,5′-dihydroxypentyl) uracil. The Journal of Biological Chemistry. 1981;**256**(19):9966-9972

[77] Davies DR, Gelinas AD, Zhang C, Rohloff JC, Carter JD, O'Connell D, et al. Unique motifs and hydrophobic interactions shape the binding of modified DNA ligands to protein targets. Proceedings of the National Academy of Sciences. 2012;**109**(49):19971-19976. DOI: 10.1073/pnas.1213933109

[78] Neubort S, Marmur J. Synthesis of the unusual DNA of *Bacillus subtilis*

bacteriophage SP 15. Journal of Virology. 1973;**12**(5):1078-1084

[79] Kirnos MD, Khudyakov IY, Alexandrushkina NI, Vanyushin BF. 2-Aminoadenine is an adenine substituting for a base in S-2L cyanophage DNA. Nature. 1977;**270**(5635):369-370

[80] Torres LL, Pinheiro VB. Xenobiotic nucleic acid (XNA) synthesis by Phi29 DNA polymerase. Current Protocols in Chemical Biology. 2018;**10**(2):e41. DOI: 10.1002/cpch.41

[81] Pinheiro VB, Taylor AI, Cozens C, Abramov M, Renders M, Zhang S, et al. Synthetic genetic polymers capable of heredity and evolution. Science. 2012;**336**(6079):341-344. Available from: http://www.sciencemag.org/content/336/6079/341%5Cnhttp://www.ncbi.nlm.nih.gov/pubmed/22517858

[82] Saier MH, Reddy BL. Holins in bacteria, eukaryotes, and archaea: Multifunctional xenologues with potential biotechnological and biomedical applications. Journal of Bacteriology. 2015;**197**(1):7-17. Available from: http://jb.asm.org/content/197/1/7.abstract

[83] São-José C. Engineering of phage-derived lytic enzymes: Improving their potential as antimicrobials. Antibiotics. 2018;7(2):29

[84] Pires DP, Cleto S, Sillankorva S, Azeredo J, Lu TK. Genetically engineered phages: A review of advances over the last decade. Microbiology and Molecular Biology Reviews. 2016;**80**(3):523-543

[85] McCafferty J, Griffiths AD, Winter G, Chiswell DJ. Phage antibodies: Filamentous phage displaying antibody variable domains. Nature. 1990;**348**:552. DOI: 10.1038/348552a0

[86] Barbas CF, Kang AS, Lerner RA, Benkovic SJ. Assembly of combinatorial antibody libraries on phage surfaces: The gene III site. Proceedings of the National Academy of Sciences. 1991;**88**(18):7978-7982. DOI: 10.1073/pnas.88.18.7978

[87] Wang HH, Isaacs FJ, Carr PA, Sun ZZ, Xu G, Forest CR, et al. Programming cells by multiplex genome engineering and accelerated evolution. Nature. 2009;**460**(7257):894-898. DOI: 10.1038/nature08187

Chapter 4

The Unusual Linear Plasmid Generating Systems of Prokaryotes

Sophie E. Knott, Sarah A. Milsom and Paul J. Rothwell

Abstract

Linear DNA is vulnerable to exonuclease degradation and suffers from genetic loss due to the end replication problem. Eukaryotes overcome these problems by locating repetitive telomere sequences at the end of each chromosome. In humans and other vertebrates this noncoding terminal sequence is repeated between hundreds and thousands of times, ensuring important genetic information is protected. In most prokaryotes, the end-replication problem is solved by utilizing circular DNA molecules as chromosomes. However, some phage and bacteria do store genetic information in linear constructs, and the ends of these structures form either invertrons or hairpin telomeres. Hairpin telomere formation is catalyzed by a protelomerase, a unique protein that modifies DNA by a two-step transesterification reaction, proceeding via a covalent protein bound intermediate. The specifics of this mechanism are largely unknown and conflicting data suggests variations occur between different systems. These proteins, and the DNA constructs they produce, have valuable applications in the biotechnology industry. They are also an essential component of some human pathogens, an increased understanding of how they operate is therefore of fundamental importance. Although this review will focus on phage encoded protelomerase, protelomerases found from *Agrobacterium* and *Borellia* will be discussed in terms of mechanism of action.

Keywords: protelomerase, telomere resolvase, linear plasmid, doggybone DNA, Touchlight Genetics

1. Introduction

The study of DNA, its structure and how it is replicated has been intensifying since the 1900s. Recent advances in DNA sequencing, bioinformatics and high-resolution imaging has increased our understanding of the variations that exist between different DNA replication systems. In general, the genetic material of bacterial cells is in the form of circular DNA molecules. Infecting bacteriophage may integrate their DNA into the host genome, or maintain it independently as a viral episome; usually this plasmid is also circular. These structures have no free ends and are therefore not susceptible to exonuclease degradation and do not suffer from the end replication problem, whereby genetic material at the tip of a chromosome is lost during each round of replication. However, some prokaryotic cells have been identified as harboring closed linear DNA chromosomes. The ends of these structures are protected by either invertron or hairpin telomeres. Invertron telomeres consist of inverted terminal repeats and covalently attached capping proteins, essential for priming DNA replication. These structures are distinct from hairpin telomeres, which have covalently closed hairpin ends.

The first linear genome of prokaryotes was obtained in 1964, when V. Ravin isolated the *Escherichia coli* phage N15 [1]. The genetic material of this phage is unusual, because it is not maintained as an independent circular entity nor is it integrated into the host genomic material. Instead, upon entry into the host, the N15 genome circularizes and is then processed into linear structures by an atypical cutting and re-joining enzyme, called a protelomerase, or telomere resolvase. Since this discovery, protelomerases have also been characterised in *Klebsiella oxytoca* phage φKO2, *Yersinia enterocolitica* phage PY54, *Vibrio parahaemolyticus* phage VP882 [2] and *Halomonas aquamarina* phage ΦHAP-1.

In addition to phage, these unusual enzymes have been isolated from certain bacteria. The best studied being ResT from *Borrelia burgdorferi* [3], the causative agent of Lyme disease. Linear chromosomes are now described as a hallmark of *Borrelia* [4] and protelomerases have been purified from *B. hermsii*, *B. parkeri*, *B. recurrentis*, *B. turicatae*, and *B. anserine* [5]. More recently, they have also been discovered in cyanobacteria [6] and the plant pathogen *Agrobacterium tumefaciens C58*, which contains a circular and linear chromosome as well as circular plasmids [7]. These proteins are clearly more widespread than initially believed and it is likely future research into other prokaryotes will identify additional members of the family.

Although currently under debate, it has been suggested protelomerases are tyrosine-recombinase-like enzymes. It remains to be determined whether the bacterial protelomerases have their origin in phage as both, to some extent, share a common substrate recognition and a DNA cleavage/rejoining mechanism. In addition, some protelomerases have been identified as having roles in vivo unrelated to telomere resolution such as single strand annealing [8] and ATP-dependent helicase activities [9]. Further research into this important class of enzymes should help elucidate the significance of single strand annealing and DNA unwinding activities during closed linear chromosome replication.

Not only are protelomerases essential for the organisms in which they reside, but their unique functionality makes them valuable to the biotechnology industry. DNA constructs produced using protelomerase are being marketed by Lucigen as improved cloning vectors for highly repetitive sequences [10, 11]. Linear structures are not susceptible to supercoiling, thus making them more stable and less susceptible to genetic loss during replication [10]. Protelomerases can also be expressed in engineered *E. coli* cells to produce linear eukaryotic vectors that contain no bacterial sequences [12]. In addition, protelomerases are a central component to Touchlight Genetics' DNA amplification platform that produces large quantities of high-quality DNA using a cell-free process, for therapeutic and industrial applications [13, 14]. Unlike plasmid DNA, the incumbent technology for therapeutic applications, Touchlight's doggybone DNA (dbDNA) platform contains no extraneous bacterial DNA sequences. The resulting minimal vector has an improved safety profile from a regulatory perspective due to elimination of antibiotic resistance genes. The small amounts of plasmid DNA required for this in vitro manufacturing process makes the dbDNA process well suited to scale production of "difficult" structured or repetitive DNA sequences or constructs that cause cell toxicity. Linear, minimal vectors may be valuable as nonviral gene therapy vectors and as DNA vaccines, both modalities gaining increasing focus and investment in the biotechnology market.

There is high interest in the study of this protein family due to their utility and potential value for biotechnology. To date, research has largely focused on characterizing protelomerase recognition sequences, solving 3-dimensional structures and exploring the effects of protein mutations on activity. An improved understanding of how protelomerases function will enhance their value for applications in synthetic biology, and may provide the opportunity to invent new and novel applications.

2. Mechanism of telomere resolution

Despite the diversity of organisms in which protelomerases reside, important features have been identified that unify and define this class of protein. The protelomerase target site, denoted as *telRL*, is a palindromic sequence of double stranded DNA. The substrate differs between protelomerases and to date only ResT, the bacterial protelomerase from *Borrelia*, has been shown to have specificity for more than one target sequence [15]. All protelomerases are thought to function as a dimer and it is widely believed that none require the addition of cofactors such as ATP or divalent cations. However, it has been shown that concentrations of EDTA >10 mm inhibit the N15 protelomerase, TelN, and the sequence of this protein predicts a binding motif for divalent cations [16].

Current models propose that protelomerases bind nonspecifically to DNA and scan the sequence until finding the target site, or coming into contact with another monomer, at which point the protein immobilizes [17]. Immobilization occurs upon dimerization, whether this forms at the substrate target sequence or not. However, only when at the correct site will the reaction of telomere formation be catalyzed. This phenomenon can be observed in vitro, where a high concentration of TelK (over 400 nm) results in the condensation of DNA and inhibition of telomere formation [17]. Protelomerase concentration in vivo therefore, must be carefully controlled. This notion has been explored in phage N15, where negative control is used to regulate the levels of protein [18].

Protelomerases catalyzes a two-step transesterification reaction, and all are thought to initiate DNA cleavage using an active site tyrosine residue. This residue performs nucleophilic attack on the phosphodiester bond to form a 3′ covalently attached protein-DNA intermediate and a free 5′-OH. The protein bound intermediate is vital for avoiding deleterious double strand breaks and prevents the premature abortion of reactions [19]. The DNA cleavage reaction happens in a staggered formation 3-bp either side of the symmetrical target site center. This leaves a 6-nucleotide overhang that loops back and is ligated to form the covalently closed hairpin end. The DNA cleaving and re-joining reactions are isoenergetic and, in principle, each step in the reaction is reversible [19]. As DNA hairpins are unable to form complete base pairings [20], they are less stable than the starting material. In this case directionality is determined by the loop processing step. This part of the reaction is poorly understood and data available indicates conflicting mechanisms in different systems.

Figure 1a is a model for telomere resolution by the protelomerase TelK from phage φKO2. An interlocked protein dimer forms at the *telRL* site and induces a sharp, roughly 73°, bend in the DNA, which displaces its helical structure and buckles the base pairs between the scissile phosphates [21]. This is described as "spring loading"; the energy stored in the distorted DNA drives the reaction forward, enabling spontaneous hairpin formation and protein dimer separation [19]. The mechanisms proposed for the bacterial protelomerases, TelA and ResT

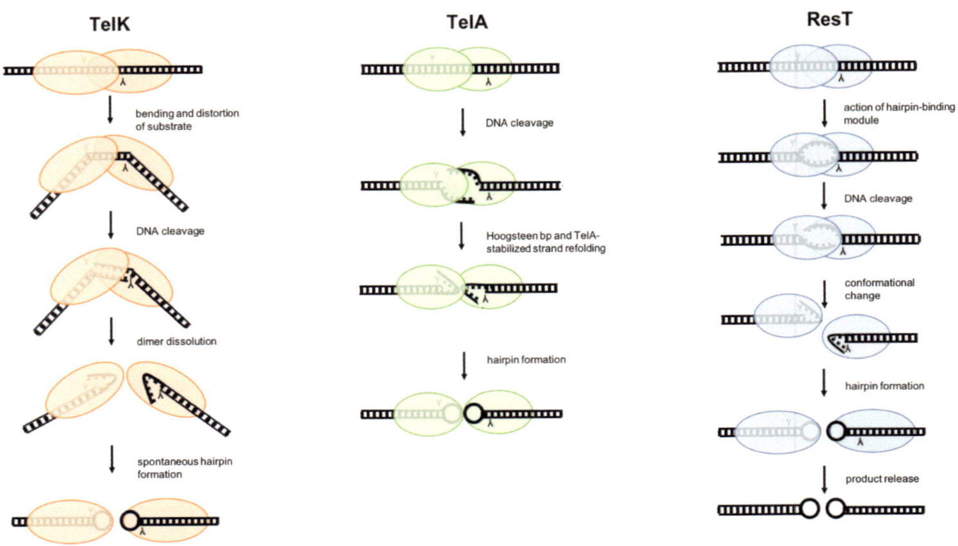

Figure 1.
Models of telomere resolution by TelK, TelA and ResT. (a) TelK monomers composed of N-terminal muzzle and C-terminal stirrup domains dimerise at the target site. This induces bending of the DNA, the spontaneous release of stored energy drives hairpin formation and dimer dissolution. (b) TelA cleaves the DNA and transient electrostatic interactions stabilize the transition state. Hairpin formation occurs within the protein dimer (c) ResT catalyzes telomere resolution with the aid of its hairpin-binding module. The final step of this reaction is product release, which is not observed for TelK or TelA [19].

(from *Agrobacterium tumefaciens C58* and *Borrelia*, respectively) are fundamentally different to that of TelK. In TelA and ResT reactions, strand refolding is enzyme-mediated, as opposed to spontaneous. A key element of the TelA mechanism is the refolding intermediate that exists before hairpin formation. This conformation is stabilized by multiple protein-DNA and DNA-DNA interactions, which drive the reaction forward by virtue of changes in binding energies. TelA binds even more strongly to the final hairpin product, thus favoring its formation. The mechanisms for TelK and TelA have been deduced from structures solved by X-ray crystallography [21, 22]. There is no structure of ResT and the mechanism proposed in **Figure 1c** is a result of research involving structure prediction, substrate modifications and protein mutations. In ResT catalyzed telomere resolution, the protein binds and distorts the DNA by underwinding at the dimer interface [19]. This is consistent with the observation that ResT has a hairpin-binding module, that presumably stabilizes the conformation of pre-hairpin DNA [23]. Hydrolysis of base pairs between the scissile phosphates promotes strand ejection following DNA cleavage. The exact mechanism of strand refolding is yet to be determined, but it is suggested to occur before dissolution of the dimer [24]. This concept of a "spring-loaded" pre-cleavage intermediate is analogous to that of TelK.

3. Substrate sequences

Identifying the natural target site of protelomerases is not straightforward. A logical strategy is to determine the nucleotide sequence of the resultant telomere and deduce from this the starting material. However, sequencing telomeres is notoriously difficult as the hairpin ends are incapable of ligating to the vector during sequencing library construction [7]. An adapted method has been used, whereby a nuclease opens the closed ends to make them compliant for ligation [25]. This does not always give absolute results, but can provide predictions for the target sequences, which may be confirmed by in vitro studies [25].

In general, protelomerases are highly specific and only process one target sequence. The exception to this is ResT, which is far less stringent and can resolve nine different telomere sequences found in the *B. burgdorferi* group of bacteria [15]. A conserved feature among all protelomerases is the palindromic nature of their substrate, with one protein molecule binding either side of the axis of symmetry to form a dimer. Interestingly, the TATAAT sequence of telomeres from N15 and φKO2 is also found in *Borrelia*. The significance of this is unconfirmed, although it has been suggested the nucleotides are important for protelomerase recognition [15]. For ResT, substitution of this sequence abolishes telomere resolution and mutating it to TTTAAT reduces the initial rate significantly [15]. Mutating the 6th and 7th nucleotide of this sequence within the TelN recognition site also produces a substrate the protein cannot process [26]. Despite functioning in different systems, TelN and TelK process highly similar target sequences, both of which are shown in **Figure 2**. These sites differ in length, but are identical in the center, and both protelomerases are capable of resolving each other's natural substrate [27]. Given the high sequence similarity (86.9%) between TelN and TelK, this observation is not hugely surprising.

Comparison of the TelN and TelK *telRL* sequences, to the 42 base pair (bp) recognition site of the PY54 protelomerase indicates limited homology and this DNA cannot be processed by any of the other protelomerases [27]. However, Huang and colleagues found altering positions 15 and 16 of the PY54 target in the top strand, plus residues 28 and 27 of the bottom strand results in a substrate that is processed, although with limited efficiency, by TelK. They went on to suggest that TelN and TelK not only recognize these specific nucleotides, but also a cruciform DNA structure that is formed [27]. Although crystal structures of TelK have since discredited the suggestion that a cruciform structure is formed [21], this work is important in that it identifies the key nucleotides that are essential for telomere resolution by these enzymes.

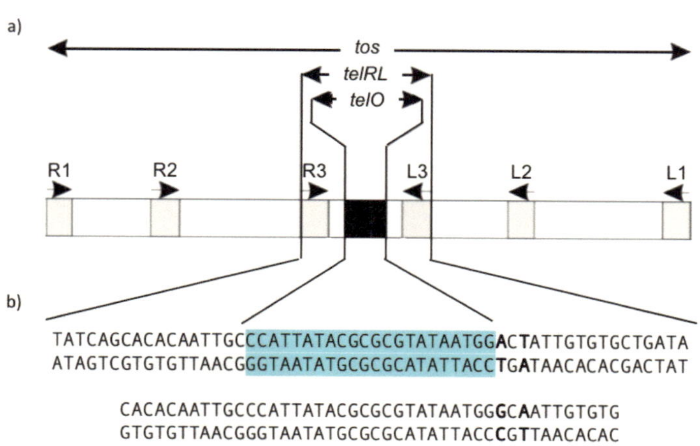

Figure 2.
Protelomerase recognition sequences. (a) The tos *site for TelN. Gray boxes indicate three regions of repeated sequences flanking the telRL site, which contains the central 22-bp telO site highlighted in cyan. Figure adapted from [16]. (b) The cognate sequence of TelN (top) and TelK (bottom). Both protelomerases are capable of processing each other's substrate. Two single point variations are shown in bold. In addition, the TelN natural substrate has six residues on each end.*

3.1 Minimal substrate

In vitro studies have also involved truncating target sites in order to identify the minimum sequence required for protelomerase binding and telomere resolution. To date, the minimal site identified that can be resolved, is a 26-bp substrate of TelA [7]. This was found by systematically deleting residues from both sides of the target sequence, until no product was produced. Similar studies have been performed on the TelN substrate. **Figure 2** shows the complete telomere occupancy (*tos*) site, which consists of a 56-bp palindromic sequence flanked by a series of inverted repeats. Initially, it was believed that *telO* is insufficient for processing by TelN, and the reaction requires the whole *telRL* site [16]. However, it has since been found that at greater TelN concentrations, roughly 50-fold higher than those required for *telRL*, the *telO* substrate is processed [26]. This indicates *telO* contains all the necessary elements for telomere resolution, but the protein requires additional sequence for binding and recognition. The binding affinity of TelN is greater still when the whole *tos* site is included in the substrate [16]. Experiments performed using ResT have explored whether the protelomerase is able to mediate cleavage on half a target site. When this half site was in a plasmid, the assay failed to produce reaction products, therefore suggesting dimer formation is essential for activity and the whole palindromic sequence is required [28].

4. Linear genomes

In order to further appreciate how protelomerases function, it is necessary to understand their role in relation to the whole phage or bacterial cell life-cycle. Bacteriophage N15 has been extensively characterised and **Figure 3** illustrates the different structures its genetic material forms upon infection of *E. coli*. The phage

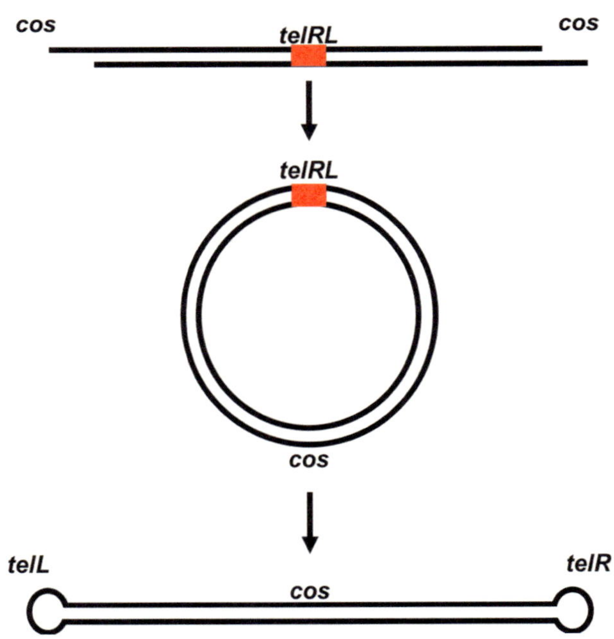

Figure 3.
Forms of bacteriophage N15 DNA. Having infected an E. coli *cell, the virion DNA circularizes, via complementary* cos *sites. Lytic or lysogenic replication can be initiated from the circular intermediate. Shown is the pathway for lysogeny, whereby the telRL is processed by protelomerases to form linear prophage DNA with covalently closed hairpin ends. Figure adapted from [1].*

DNA is a 46.4 kb chromosome that has two cohesive end sites (*cos*) consisting of 12-nucleotide overhangs at each 5′-end. These sites are complementary and can be ligated to form circular DNA. This circular intermediate then acts as the starting material for either lytic (not shown in **Figure 3**) or lysogenic development. During lysogenic development the *telRL* site is recognized and processed by protelomerases, this reaction forms a linear DNA structure with covalently closed ends. A similar genome arrangement has been identified for φKO2 [27], VP58.5 [29], VP882 [2] and PY54 [30]. These prophages all have cohesive (*cos*) ends that presumably enable the formation of similar structures as those described for N15. Interestingly, no *cos* site has been identified in the ΦHAP1 genome [31], therefore indicating a different mechanism of DNA packing. The discovery of terminase genes [31], suggests that headful packing may occur, whereby concatermeric DNA is packed into the phage capsid until it is full [32].

5. Bacteriophage N15 replication

Various models have been proposed to describe the replication and processing of linear DNA with hairpin telomeres [33]. Uncertainties arise about the specific mode of replication and whether it occurs uni- or bi-directionally. Other important factors that need to be determined are, where in the plasmid replication is initiated from and what the replication intermediates are. Bacteriophage N15 can be used as the model system to explore these questions, the general organization of its genome is shown in **Figure 4**. Genes have been largely identified by homology inferred from sequence similarity to other bacteriophages; mainly lambda, HK97 and HK002 [1]. The division between the left- and right-hand side of the N15 genome is marked by *telRL*. The left arm encodes structural proteins required for N15 head and tail assembly. The right-arm contains more unusual genes and only 10 of the 35 have identified homologs in other lambdoid phage [1]. These are therefore much harder to characterize, and it is yet to be determined how they all function during N15 replication.

RepA is the only gene essential for replication of prophage N15 DNA [18]. It encodes a large, multifunctional protein that has both primase and helicase activities [35]. Sequence alignments have highlighted regions of RepA with similarities to both plasmid and viral DNA replication proteins [1]. Most notably, the phage P4 alpha protein [1] also has combined primase and helicase activities [36]. Phage P4 replication occurs by a theta-mechanism [37]. The similarities between alpha protein and RepA, combined with studies measuring amplification rates of DNA markers [18], strongly suggests that typical bidirectional theta-replication also occurs in N15 prophage. The origin of replication (*ori*) resides within the *repA* gene [18], which is located closer to the left hairpin end of the plasmid.

The gene *telN* encodes a 71 kDa protein that has partial homology to integrases and an amino acid sequence characteristic of those that bind DNA as homo- or hetero- dimers [38, 39]. This was correctly identified as the protelomerase encoding

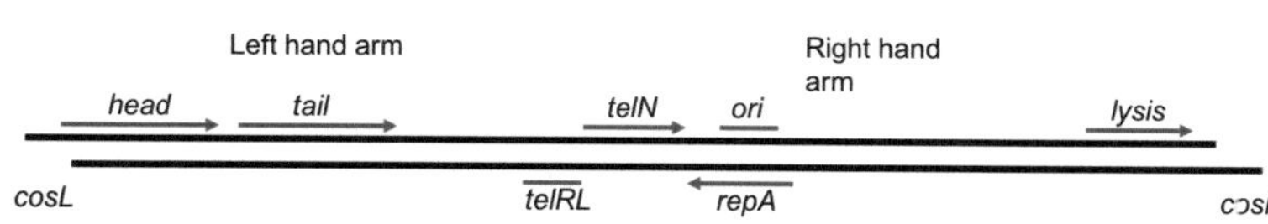

Figure 4.
The chromosome of bacteriophage N15. 46.4 kb double stranded DNA with 12-bp single stranded cohesive termini (cosL and cosR). Arrows indicate the direction of transcription and the telRL site divides the sequence into left- and right-hand arms. Adapted from [34].

gene and in 2000, Deneke and colleagues purified its protein product [16]. TelN is capable of processing the 56-bp *telRL* site in both linear and circular supercoiled DNA [16]. To decipher the mechanism of N15 genomic replication, mutants deficient in this protein have been created [40]. In protelomerase-deficient cells, unprocessed replicative intermediates accumulate, the structures of which have been characterised as circular head-to-head dimer molecules [40].

Figure 5 describes how these linear N15 constructs may be replicated and processed; it is consistent with the data cited above and proposes structures that have been validated by electron microscopy [18]. In pathway A, following replication of the *telL* site, TelN processes the DNA to create a Y-shaped structure. After duplication of *telR*, the right telomere is also modified to form the final linear product. Alternatively, in pathway B the whole DNA molecule has been replicated, producing a head-to-head circular dimer that is then resolved. Interestingly, this mechanism of replication is distinct from that described for eukaryotic replicons, even those with similar telomeric ends, therefore suggesting an independent evolution [18].

5.1 Lytic replication

A model of how N15 lytic replication could occur is proposed in pathway C of **Figure 2**. The DNA is duplicated and resolved into two circular monomers, as opposed to linear structures. These circular molecules are the starting material for subsequent cycles of amplification. This style of lytic replication is similar to that of phage lambda. This bacteriophage also circularizes its DNA upon entering the host cell, it has *cos* sites analogous to those of N15 [1]. Further similarities between N15 and lambda include: genome length, burst size, latent period, lysogenization frequency and phage particle and plaque morphology [38]. Their structural and packing proteins are also highly analogous, making it likely that N15 DNA packing follows a pathway similar to that of lambda [1]. It has even

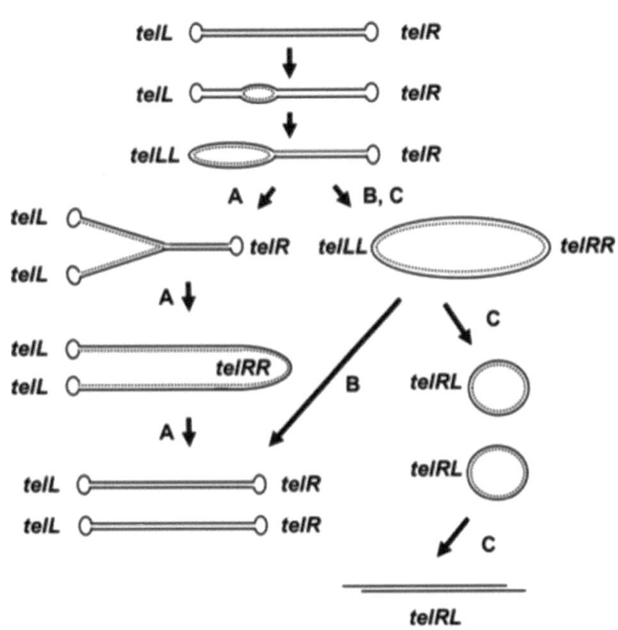

Figure 5.
Model for lytic and lysogenic replication of N15 linear prophage. Bi-directional theta-replication begins at the internal ori *site. A: Duplicated* telL *sites are processed before complete replication of genome. B: The template is completely duplicated prior to processing by TelN. C: Lytic replication, whereby circular monomers are produced that then undergo subsequent rounds of replication [18].*

been demonstrated that the N15 specific terminase can package lambda DNA with reasonable efficiency [41].

The key difference between N15 and lambda bacteriophage is that lambda integrates its DNA into the host genome, whereas N15 does not. Although protelomerases share some sequence homology with lambda integrases, and both appear to have comparable roles in helping establish prophage DNA, these proteins are not functional analogues. During lytic replication the lambda integrase is dispensable, in comparison the protelomerase of N15 is essential. This phenomenon has been proven by experiments showing N15 deficient in protelomerases are incapable of infecting *E. coli* cells [35], although why this is the case remains unclear. As the establishment of lytic growth requires the conversion of linear plasmid molecules to circular ones, it could be presumed a protelomerase mediated “telomere fusion” reaction occurs. However, TelK, which is highly analogous to TelN, is incapable of catalyzing this in vitro [21], it is therefore highly unlikely wild-type TelN is functioning in this way. The possibility of an unknown factor modifying the protelomerase and/or its target site to prevent the usual processing reaction cannot be ruled out. In lambda, Xis, assisted by the host factor Fis, is necessary to induce excision during induction of a lysogeny [42]. Potentially, an analogue could be encoded by one of the N15 late genes [35], although experimental evidence to support this theory is yet to be provided. However, it has been demonstrated that mutating histidine 415 of TelN to an alanine results in accumulation of circular head-to-tail monomers [35]. This histidine is important for catalytic activity and is believed to coordinate the scissile phosphate [16]. Interestingly, its mutation does not have the same effect as mutating the catalytic tyrosine 424, which acts as the nucleophile in telomere resolution. When this residue is changed to an alanine, accumulation of circular dimers does occur, but in this case, they are “head-to-head” as opposed to “head-to-tail” [35]. The significance of this observation is currently unknown. Given that TelN cannot be recycled [27], it has also been proposed that the protein’s depletion will result in fewer linear molecules being produced and a natural accumulation of head-to-head dimers, which can then be processed to circular structures [35].

5.2 N15 as a model for the replication of other linear plasmids

To what extent can the model of bacteriophage N15 be extended to describe the replication of other replicons with hairpin ends? Genomic sequence analysis of phage encoded protelomerases reveal little overall sequence similarity [43]. However, the organization of functional domains is analogous and they appear to have conserved regulatory regions [43]. This would suggest a shared mechanism of plasmid replication and lysogeny control [43]. Virions of φKO2, VP58.5, VP882 and PY54 have cohesive ends [27, 29, 2], which facilitate circularization and enable the formation of similar structures as those described for N15. The absence of *cos* sites in the ΦHAP1 genome has already been discussed and suggests a different mechanism of DNA packing [31]. Importantly, these phages all have homologs of the N15 protelomerase and replication protein RepA. The genes encoding these proteins are found between the lysogeny control region and structural gene cluster, as is the case is N15 [43]. Although yet to be confirmed by in vitro studies, given these similarities it is sensible to suggest that replication of these linear phage plasmids follows a model comparable to that proposed for N15.

Further comparisons can be made between the suggested phage model and that of bacterial linear chromosome replication. *B. burgdorferi* has a linear chromosome [44], and it is replicated in a bi-directional manner to produce circular, head-to-head intermediates [45], which are then processed by telomere resolution [46]. Here replication is also initiated at an internal *ori* site and the protelomerase, ResT, is known to be essential [47]. These findings indicate a shared fundamental mechanism of genomic replication between N15 and *Borrelia*. Nonetheless, discrepancies between the different systems have been highlighted. For one, in these bacterial cells the protelomerase is encoded, not on the same DNA construct it processes, but on a different circular plasmid, cp26 [3]. In addition, the possibility of *Borrelia* accessory factors influencing telomere resolution was suggested, following the observation that differential processing occurs in vitro compared to in vivo [15]. Potentially this could be a result of in vitro conditions not completely reconstructing those occurring in vivo. However, if correct it would indicate important discrepancies between how bacterial and phage protelomerases are regulated.

6. Structural data

6.1 X-ray crystallography

The X-ray structures described for both TelK and TelA have greatly enhanced our understanding of the protelomerase mechanism [21, 22]. The structure of TelK (from phage φKO2), is shown in **Figure 6**; crystallized in a dimer conformation

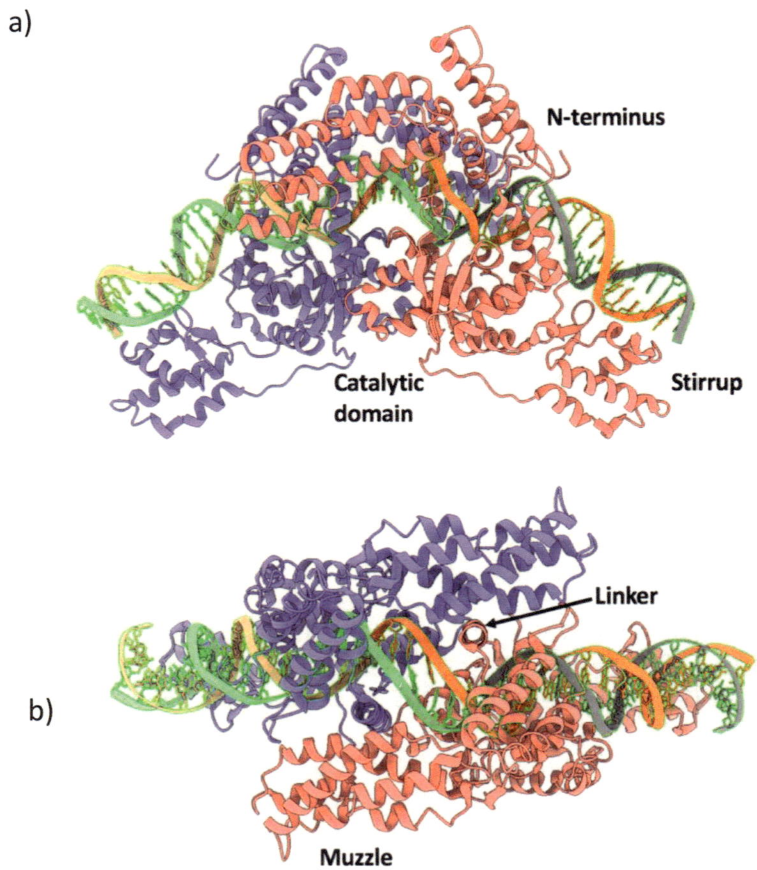

Figure 6.
Crystal structure of TelK dimer complexed with DNA. (a) Two monomers of TelK dimers at the recognition site and are held together by multiple transient protein-protein and protein-DNA interactions. The structure of each monomer is largely alpha helical, with mixed beta strands at the catalytic domain. (b) The same complex viewed from the N-terminus. The helical linker fits in the major groove of the DNA, it contacts the DNA on the opposite side to the rest of the protein. The structures presented were generated with using ChimeraX (Goddard et al., 2018) PDB ID: 2v6e [21].

complexed with the minimal cognate DNA sequence of 44-bp [PDB: 2V6E]. TelK has been divided into three core domains, all of which make contact with the DNA. These include, the muzzle at the N-terminus, the catalytic domain in the center, and the stirrup domain at the C-terminus. A long alpha-helical linker is also highlighted, and this connects the core catalytic and N-terminal domains.

One of the most striking observations of this structure is the level distortion: DNA is bent at roughly 73° parallel to the axis of symmetry [21]. This provides a valuable insight into the mechanism of telomere resolution by TelK and would appear to refute previous theories that the DNA is forced into a cruciform conformation [27]. Core substrate binding occurs at the N-terminus and the muzzle makes

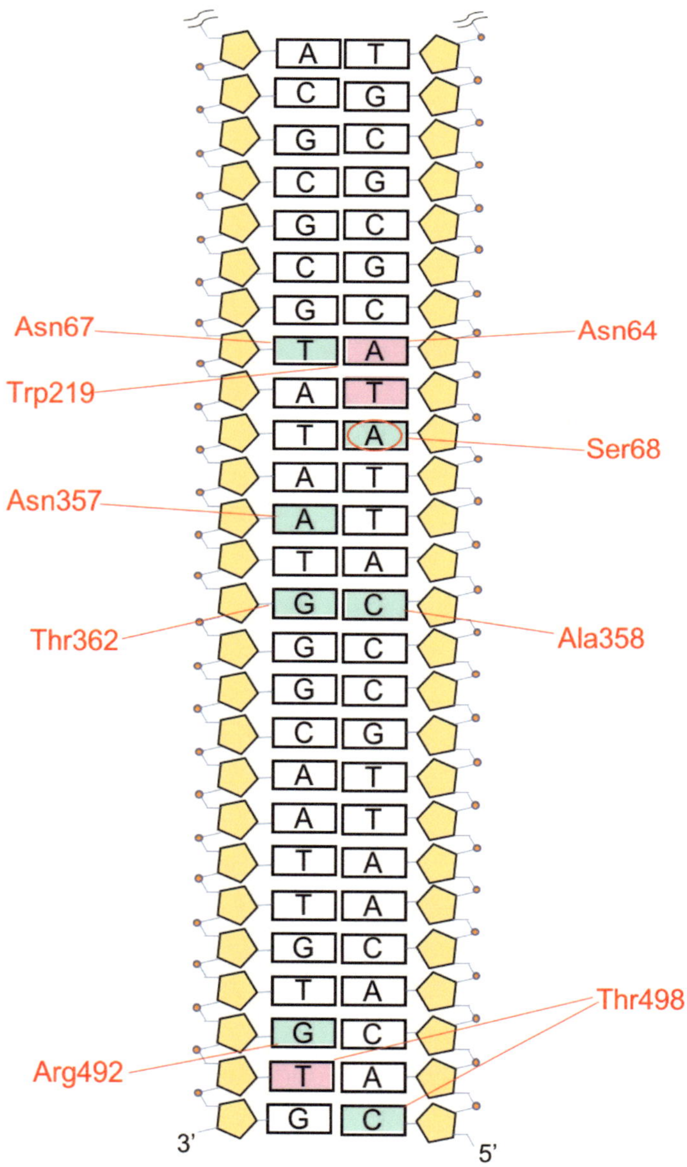

Figure 7.
TelK substrate recognition. One half of the TelK recognition sequence, nucleotides forming hydrogen bonds to protein are highlighted in cyan and those forming van der Waals interactions, in pink. Multiple amino acids interact with the DNA backbone and are not shown, those forming bonds with the bases are illustrated in red. Adenine at position 42 is circled, this residue forms hydrogen bonds with serine 68 of TelK and is therefore important for site recognition. Adapted from [21].

extensive contacts with the opposite subunit, strengthening the protein's structure and pushing the DNA into this strained conformation. The catalytic site is formed at the dimer interface, it binds to the opposite side of the substrate relative to the N-terminus and the helical linker that connects these two domains is fixed in the major groove. Extensive electrostatic interactions mediate the interaction.

Closer examination of the interactions between the DNA and protein reveal seven nucleotides that form hydrogen bonds with nearby residues (shown in cyan in **Figure** 7) and are presumed key for substrate recognition. This model is supported by the previously cited studies, whereby the natural PY54 substrate was effectively mutated into a sequence that could be processed by TelK. Adenine at position 42 is circled; this is one of the bases identified as forming hydrogen bonds with TelK and is one of the points that required mutating in order to form a cognate sequence. High salt has been shown to inhibit telomere resolution by protelomerases [16], this is possibly reflected by the extensive hydrophilic protein-DNA interactions that would be disrupted by an excess of ions.

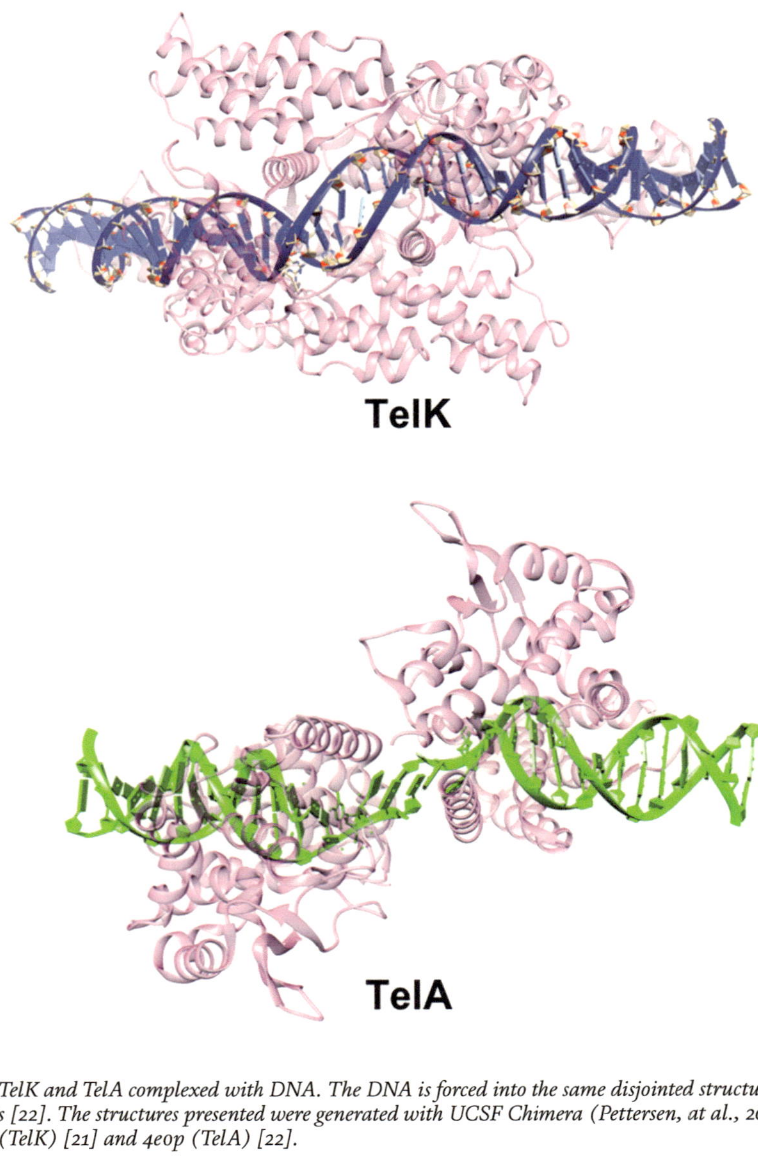

Figure 8.
Dimers of TelK and TelA complexed with DNA. The DNA is forced into the same disjointed structure down its helical axis [22]. The structures presented were generated with UCSF Chimera (Pettersen, at al., 2004) using PDB 2v6e (TelK) [21] and 4e0p (TelA) [22].

The stirrup domain of TelK has a winged helix-turn-helix motif [21], it makes few contacts with the rest of the protein but extends the DNA binding interface. In stabilizing the strained substrate conformation, this part of the protein aids hairpin formation; however, it is nonessential for the cleavage reaction [21]. Following strand cleavage, the stored energy is released, and this drives dimer dissolution which is proceeded by spontaneous hairpin formation. The stirrup is not conserved among protelomerases and this provides further evidence to support the theory that the mechanism of telomere resolution varies between different systems.

The structure of the bacterial protelomerase, TelA has also been described via X-ray crystallography. TelA is considerably smaller than TelK, it lacks the stirrup and only consists of the catalytic and N-terminal domains. This design is similar to that of tyrosine recombinases, which are also typically composed of two domains [48]. The N-terminal 100 residues are poorly resolved in comparison to the rest of the protein, this area of low electron density suggests flexibility of the polypeptide chain. Comparing dimer substrate complexes of TelA to those of TelK (**Figure 8**) reveals a similar DNA conformation at the dimer interface. Extensive hydrogen bonds and van der Waals interactions are also involved in dictating the substrate specificity of TelA and the DNA exhibits the same disjunction down its helical axis [22].

6.2 Catalytic domain

The catalytic domain of TelK is a mixed alpha beta structure. **Figure 9** shows the core catalytic residues of TelK R275, K300, K380, R383 and H416. These act together to maintain catalytic activity and coordinate nucleophilic attack of the tyrosine. Side chains of the basic amino acids at positions one, three and four provide a hydrogen bonding network that coordinates the scissile phosphate and stabilizes the transition state [22]. In type IB topoisomerases, the pentad is usually composed of RKKRH/N [49], with basic residues 1 and 3 having the same stabilization effect. In these proteins, the second lysine residue has been shown to donate a proton to the

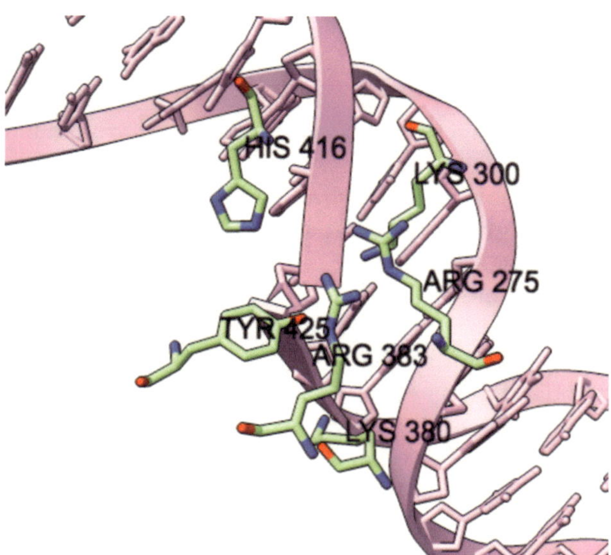

Figure 9.
Conformation of protelomerase active site residues. The residues of TelK, R275, K300, K380, R383 and H416 maintain the active site conformation and coordinate nucleophilic attack of the tyrosine. The structures presented were generated with ChimeraX (Goddard, T et al., 2018) using PDB 2v6e [21].

5′-OH leaving group, which aids its removal during cleavage [50]. It is feasible that the lysine in the protelomerase's active site, with its side chain positioned between the DNA O5' and nonbridging oxygen, also functions in this way and protonates the leaving group [21].

These crystal structures are invaluable when trying to decipher the mechanism of protelomerase-catalyzed telomere resolution. TelN and TelK have highly homologous sequences and can process the same target site [27], it is therefore likely the structures of these enzymes are analogous and information about TelN can be derived from examining the TelK structure. However, is important to note that the crystallized structure of TelK is not full length and lacks 100 residues from the C-terminus. This does not appear to significantly affect protein activity in vitro [21], but the significance in vivo is unconfirmed.

7. Evolutionary history

It is widely believed Tyrosine recombinases and type IB topoisomerases have arisen from a common ancestor [50]. These structurally and mechanically analogous proteins catalyze important DNA rearrangements, they use an active site tyrosine residue and form covalently bound protein-DNA intermediates. Similarities between this mechanism, and that of protelomerases, has led to the suggestion that these protein classes have a shared evolutionary relationship [51].

Key to the function of tyrosine recombinases is the conserved catalytic motif: "RKHRH" [52, 53]. Crystal structures and sequence alignments reveal that protelomerases share this catalytic pentad [27], with exception of the middle residue, which varies between systems. Deviation at this central point is also observed in tyrosine recombinases, where the central histidine may be replaced by an arginine, asparagine, lysine or tyrosine. In protelomerases; TelN and TelK both have a methionine at this position, PY54 contains a lysine, VHML has a histidine and the bacterial protelomerases of *B. burgdorferi* and *A. tumefaciens*, have a tyrosine. It has been found substituting the tyrosine to a histidine or lysine in TelA is completely tolerable and results in no loss of activity [7]. Whether these variations are of significance is yet to be determined.

Structural comparisons outside of the catalytic domain reveal low overall sequence homology at both the N- and C- terminus. ResT is smaller than the phage proteins, partial proteolysis separates the 449 amino acid protein into two domains [54], and sequence analysis predicts its architecture is more comparable to that typical of tyrosine recombinases. In addition, this protein has the unusual and surprising ability to synapse Holliday Junctions [51]. The reaction appears to be favored in conditions that are counterproductive to telomere resolution, such as negative supercoiling, or unsymmetrical *telR* sites [51]. This observation provides compelling evidence for the argument that ResT may have evolved from a recombinase [51]. Although the significance of Holliday Junction formation by ResT is still under investigation, it has been suggested this could be an obsolete ancestral property of the protein [4]. Additional evidence highlighting a relationship between tyrosine recombinases and protelomerases comes from the Flp recombinase and lambda integrase. It has been found that, under specific conditions, these enzymes have the ability to form hairpin products [55, 56]. Thus, indicating the relative ease with which recombinases may be converted to telomere resolving enzymes.

In ResT, additional catalytic residues outside of the core pentad are required for telomere resolution [57]. This indicates digression from a typical tyrosine recombinase type mechanism. Furthermore, mutation of ResT histidine 324, the fifth residue of the catalytic motif does not result in total loss of activity [57]. This would mark an obvious disparity between these two classes of enzymes, if it was not for the observation that this residue is also not essential in Flp recombinases [58]. Here it appears the final amino acid has a structural rather than catalytic role [58], which could be mirrored in ResT.

Analysis of the DNA substrates reveal further fundamental differences between protelomerases and tyrosine recombinases. Aside from ResT, which it has been suggested can act on more than one target site, protelomerases are specific and have stringent substrate sequence requirements. The target site of one recombinase often includes many other related points, such as phage or bacterial attachment sites. Furthermore, the reaction is intramolecular and can require auxiliary proteins [59]. This implies a mechanism more involved and complex than that of telomere resolution. The reaction catalyzed by type IB topoisomerases is considerably simpler, as these proteins act as monomers [60] and do not display the same sequence specificity as protelomerases.

In conclusion, there are key differences between the catalytic mechanism of protelomerases and tyrosine recombinases/type IB topoisomerases. However, there are also significant similarities and whether these proteins have evolved from a common ancestor is difficult to determine. It is expected that the conversion of a tyrosine recombinase to an enzyme capable of telomere resolution would be accompanied by the linearization of plasmid DNA [4]. The data suggesting this could be achieved with relative ease adds support to the argument these proteins are related. Under certain conditions, recombinases can have topoisomerase activity, and topoisomerases can affect DNA strand exchanges [27]. This raises an interesting question as to whether protelomerases, under the correct conditions, may also be able to exhibit topoisomerase activity [27].

8. Applications in biotechnology

The unique properties of protelomerases, and the DNA structures they produce, makes them a valuable class of protein that have important applications in synthetic biology and biotechnology. Linear DNA does not exhibit the supercoiling associated with plasmids, as the ends are free to rotate. This makes them stable vectors invaluable for cloning difficult sequences. DNA that is rich in adenine and thymidine, or contains lots of short tandem repeats is typically hard, if not impossible, to clone into circular vectors [10]. A commercially available cloning vector, pJAZZ from Lucigen, is based on the linear N15 phage genome. pJAZZ is sold as part of a cloning system (BigEasy Kit), enabling the insertion of otherwise unclonable sequences into the vector for the creation of viable linear plasmids that can be transformed into cells. pJAZZ vectors encode RepA and TelN, essential for bidirectional replication and telomere resolution. Transcriptional terminators flank the cloning site, thus minimizing interference and preventing transcription into and out of this region. These modifications extend the cloning possibilities and allow for the insertion of large cDNAs or operons [10].

The pJAZZ system has been further modified to specifically enhance its efficiency for the production of in vitro transcribed (IVT) mRNA. IVT mRNA is a powerful therapeutic tool, enabling the transient expression of heterologous proteins [61, 62]. However, in order to optimize translation efficiency and mRNA stability, the poly(A) tail length of the mRNA needs to be defined and optimized [63]. In particular, mRNA with poly(A) tails >300 nucleotides and purines at the 3′ end have been demonstrated as highly effective and appear to exhibit improved translation properties compared to those with shorter poly(A) tracts [64]. Due to its linear structure, a pJAZZ derived plasmid, called p(Extended Variable Length) (pEVL), can create poly(A) tracts of up to 500 bps in length. Furthermore, the residues at the 3′ end can be defined as either adenine or guanine. This has a significant advantage over conventional circular DNA, which cannot incorporate more than 174 bp of poly(A) tract without conferring extreme instability [64].

Mediphage Bioceuticals is a genetic medicine company that has also utilized the unique DNA processing capability of protelomerases for their technology. They have developed a one-step in vivo platform to produce linear covalently closed constructs, called ministring DNA [65]. These constructs are produced in *E. coli* cells that have been engineered to express the PY54 protelomerase under the control of a heat-inducible promoter. Following induction, the protelomerase processes precursor plasmids in the cell, in doing so it effectively separates the desired expression cassette from the bacterial plasmid backbone. The ministring DNA can then be purified and used as a vector for gene or cell therapies and gene editing. Although ministring production is reliant on large scale bacterial fermentation, the absence of bacterial DNA elements render these constructs preferential to plasmid DNA for medicinal applications. Furthermore, the constructs are typically smaller than plasmids, this enhances their transfection efficiency thus making them less toxic, as fewer transfection reagents are required [12]. They are also more resistant to the shear-induced degradation that large plasmids are highly susceptible to [66].

Another in vivo application of protelomerases has been explored by Katzen at colleagues, who used TelN to fragment an *E. coli* chromosome into smaller, autonomous units [67]. These proof of concept experiments were designed as a solution to the considerable difficulties associated with synthesizing and manipulating large, stable genetic elements. In splitting the chromosome into two smaller units, which together contain the essential components required for cell viability, they significantly simplified the problem. Not only are the smaller units of genetic material easier to manipulate, but each episome is of a size that it can be assembled without the need for an assembly host. This work may be extended to fragment and linearize other genomic elements of interest, in particular for the study of large units, >2 Mbp, which at present cannot be assembled and maintained in any biological platform [67].

Touchlight Genetics Ltd. has also utilized protelomerases for their technology. The platform they have developed is a purely in vitro DNA production process that eliminates all the major problems associated with using bacterial fermentation for DNA amplification. Their cell-free technology uses a phage DNA polymerase from Phi29 to produce large amounts of DNA concatemers from small amounts of starting plasmid DNA template. DNA concatemers are processed by the protelomerase TelN, to create linear covalently closed constructs that are marketed as dbDNA. The closed ended linear DNA construct is able to encode long, difficult DNA sequences which are not tolerated in high yield production of plasmid DNA due to selection pressures. The in vitro amplification technology produces DNA containing no bacterial origin of replication sequences or antibiotic selection marker. These vectors are capable of immunizing against influenza infection, with a response comparable to that of plasmid DNA [68], as well as improving tumor growth control in a Human papillomavirus (HPV) driven head and neck cancer model by delivering a therapeutic vaccine encoding for HPV16 E6 and E7 antigens [69]. Furthermore, dbDNA constructs have been used to generate functional lentiviral vectors [14] and are promising candidates for the production of recombinant adeno-associated virus (AAV) vectors [13]. This platform could therefore be important for gene therapy [70, 71] and this synthetic process for DNA production will have a broad range of application within the wider synthetic biology field.

9. Conclusion

Protelomerases are an interesting and unique class of protein. In forming telomeric structures at the ends of linear plasmids, they protect the genetic material from degradation and provide a novel solution to the end replication problem. They have also been implemented as an essential component of certain human pathogens and have important applications in synthetic biology. Despite this, protelomerases remain poorly characterised and many questions about their structure, function, mechanism and evolutionary history remain unanswered.

Much of our current knowledge has been obtained by combining information from crystal structures, analysing sequences and performing in vitro assays with protein and substrate variations. The results of these studies have enabled us to compare the protelomerases from different systems and it is clear that there is much variety within this protein family. In particular, ResT has been identified as having additional, largely unexplainable, functionality aside from telomere resolution. Potentially further characterization of the other protelomerases will lead to similar revelations.

Biochemical analysis of protelomerases from ΦHAP1, PY54 and VP58.2, will shed further light on the underlying properties that differentiate these telomere resolving enzymes. Such work could also explore the evolution of protelomerases and prokaryotes with linear plasmids. Similarities in the genome organization of telomere phage suggest a common ancestor [43] and whether the bacterial proteins originated from these is currently unknown. Introducing different phage into the same cell and determining their compatibility can give insight into evolutionary background [72]. An understanding of the relationship between phage lambda and telomere phage, in particular N15, will provide an interesting and important insight into how plasmid and phage can interact and evolve.

Crystal structures of TelK and TelA provide a solid starting point for research aiming to solve structures of protelomerases. These can also be used for homology modeling to infer information about proteins with high sequence similarities. Advances in the field of synthetic biology and protein engineering make detailed knowledge of these proteins even more valuable. An increased understanding of how they operate, and which parts are responsible for specific functionalities, will open up opportunities to produce variants with altered activities.

Appendices and nomenclature

dbDNA	doggybone DNA
bp	base pair
Tos	Telomerase occupancy site
Cos	cohesive end sites
Ori	origin of replication
IVT	in vitro transcribed
pEVL	p(Extended Variable Length)
AAV	adeno-associated virus
HPV	Human papillomavirus

Conflict of interest

SK has recently started a PhD jointly between Touchlight Genetics and Renos Savva; however, this has not relatively biased or affected the content of this chapter.

Author details

Sophie E. Knott, Sarah A. Milsom and Paul J. Rothwell*
Touchlight Genetics, Hampton, UK

*Address all correspondence to: paul.rothwell@touchlight.com

References

[1] Ravin V, Ravin N, Casjens S, Ford ME, Hatfull GF, Hendrix RW. Genomic sequence and analysis of the atypical temperate bacteriophage N15. Journal of Molecular Biology. 2000;**299**(1):53-73

[2] Lan SF, Huang CH, Chang CH, Liao WC, Lin IH, Jian WN, et al. Characterization of a new plasmid-like prophage in a pandemic *Vibrio parahaemolyticus* O3:K6 strain. Applied and Environmental Microbiology. 2009;**75**(9):2659-2667

[3] Kobryn K, Chaconas G. ResT, a telomere resolvase encoded by the Lyme disease spirochete. Molecular Cell. 2002;**9**:195-201

[4] Chaconas G, Kobryn K. Structure, function, and evolution of linear replicons in *Borrelia*. Annual Review of Microbiology. 2010;**64**(1):185-202

[5] Kobryn K, Chaconas G. Hairpin telomere resolvases. Microbiology Spectrum. 2014;**2**(6):273-287

[6] Loh T, Elvitigala T, Wang C, Wollam A, Welsh EA, Liberton M, et al. The genome of Cyanothece 51142, a unicellular diazotrophic cyanobacterium important in the marine nitrogen cycle. Proceedings of the National Academy of Sciences of the United States of America. 2008;**75**(9):2659-2667

[7] Huang WM, DaGloria J, Fox H, Ruan Q, Tillou J, Shi K, et al. Linear chromosome-generating system of *Agrobacterium tumefaciens* C58: Protelomerase generates and protects hairpin ends. The Journal of Biological Chemistry. 2012;**287**(30):25551-25563

[8] Mir T, Huang SH, Kobryn K. The telomere resolvase of the Lyme disease spirochete, *Borrelia burgdorferi*, promotes DNA single-strand annealing and strand exchange. Proceedings of the National Academy of Sciences of the United States of America. 2013;**41**(22):10438-10448

[9] Huang SH, Cozart MR, Hart MA, Kobryn K. The *Borrelia burgdorferi* telomere resolvase, ResT, possesses ATP-dependent DNA unwinding activity. Nucleic Acids Research. 2017;**45**(3):1319-1329

[10] Godiska R, Mead D, Dhodda V, Wu C, Hochstein R, Karsi A, et al. Linear plasmid vector for cloning of repetitive or unstable sequences in *Escherichia coli*. Nucleic Acids Research. 2010;**38**(6):e88

[11] Godiska R, Ravin N, Mead D. Linear plasmid vector for cloning "Unclonable" DNA. BioTechniques. 2008;**45**(5):592

[12] Nafissi N, Slavcev R. Construction and characterization of an in-vivo linear covalently closed DNA vector production system. Microbial Cell Factories. 2012;**11**:154. DOI: 10.1186/1475-2859-11-154

[13] Karbowniczek K, Rothwell P, Extance J, Milsom S, Lukashchuk V, Bowes K, et al. Doggybone™ DNA: An advanced platform for AAV production. Cell & Gene Therapy Insights. 2017:731-738

[14] Karda R, Counsell JR, Karbowniczek K, Caproni LJ, Tite JP, Waddington SN. Production of lentiviral vectors using novel, enzymatically produced, linear DNA. Gene Therapy. 2019;**26**:86-92

[15] Tourand Y, Kobryn K, Chaconas G. Sequence-specific recognition but position-dependent cleavage of two distinct telomeres by the *Borrelia burgdorferi* telomere resolvase, ResT. Molecular Microbiology. 2003;**48**:901-911

[16] Deneke J, Ziegelin G, Lurz R, Lanka E. The protelomerase of temperate *Escherichia coli* phage N15 has cleaving-joining activity. Proceedings of the National Academy of Sciences of the United States of America. 2000;**97**(14):7721-7726

[17] Landry MP, Zou X, Wang L, Huang WM, Schulten K, Chemla YR. DNA target sequence identification mechanism for dimer-active protein complexes. Nucleic Acids Research. 2013;**41**(4):2416-2427

[18] Ravin NV. Mechanisms of replication and telomere resolution of the linear plasmid prophage N15. FEMS Microbiology Letters. 2003;**221**(1):1-6

[19] Lucyshyn D, Huang SH, Kobryn K. Spring loading a pre-cleavage intermediate for hairpin telomere formation. Nucleic Acids Research. 2015;**43**(12):6062-6074

[20] Chou S, Chin K, Wang AH. Unusual DNA duplex and hairpin motifs. Nucleic Acids Research. 2003;**31**(10):2461-2474

[21] Aihara H, Huang WM, Ellenberger T. An interlocked dimer of the protelomerase TelK distorts DNA structure for the formation of hairpin telomeres. Molecular Cell. 2007;**27**(6):901-913

[22] Shi K, Huang WM, Aihara H. An enzyme-catalyzed multistep DNA refolding mechanism in hairpin telomere formation. PLoS Biology. 2013;**11**(1):e1001472

[23] Bankhead T, Chaconas G. Mixing active-site components: A recipe for the unique enzymatic activity of a telomere resolvase. Proceedings of the National Academy of Sciences. 2004;**101**(38):13768-13773

[24] Briffotaux J, Kobryn K. Preventing broken *Borrelia* telomeres ResT couples dual hairpin telomere formation with product release. The Journal of Biological Chemistry. 2010;**285**(52):41010-41018

[25] Casjens S, Murphy M, Sampson L. Telomeres of the linear chromosomes of Lyme disease spirochaetes: Nucleotide sequence and possible exchange with linear plasmid telomeres. Molecular Microbiology. 1997;**26**:581-596

[26] Deneke J, Lurz R, Lanka E. Phage N15 telomere resolution. The Journal of Biological Chemistry. 2002;**277**(12):10410-10419

[27] Huang WM, Joss L, Hsieh T, Casjens S. Protelomerase uses a topoisomerase IB/Y-recombinase type mechanism to generate DNA hairpin ends. Journal of Molecular Biology. 2004;**337**(1):77-92

[28] Kobryn K, Burgin AB, Chaconas G. Uncoupling the chemical steps of telomere resolution by ResT. The Journal of Biological Chemistry. 2005;**280**(29):26788-26795

[29] Zabala B, Hammerl JA, Espejo RT, Hertwig S. The linear plasmid prophage Vp58. 5 of *Vibrio parahaemolyticus* is closely related to the integrating phage VHML and constitutes a new incompatibility group of telomere phages. Journal of Virology. 2009;**83**(18):9313-9320

[30] Hertwig S, Klein I, Lurz R, Lanka E, Appel B. PY54, a linear plasmid prophage of *Yersinia enterocolitica* with covalently closed ends. Molecular Microbiology. 2003;**48**(4):989-1003

[31] Mobberley JM, Authement RN, Segall AM, Paul JH. The temperate marine phage HAP-1 of *Halomonas aquamarina* possesses a linear plasmid-like prophage genome. Journal of Virology. 2008;**82**(13):6618-6630

[32] Black L. DNA packaging in dsDNA. Annual Review of Microbiology. 1989;**43**:267-292

[33] Casjens S. Evolution of the linear DNA replicons of the *Borrelia* spirochetes. Molecular Microbiology. 1999;**2**(5):529-534

[34] Ravin NV. N15: The linear phage-plasmid. Plasmid. 2011;**65**(2):102-109

[35] Mardanov AV, Ravin NV. Conversion of linear DNA with hairpin telomeres into a circular molecule in the course of phage N15 lytic replication. Journal of Molecular Biology. 2009;**391**(2):261-268

[36] Ziegelin G, Scherzinger E, Lurz R, Lanka E. Phage P4 alpha protein is multifunctional with origin recognition, helicase and primase activities. The EMBO Journal. 1993;**12**(9):3703-3708

[37] Orejas RD, Ziegelin G, Lurz R, Lanka E, Genetik MM, Schuster A. Phage P4 DNA replication in vitro. Nucleic Acids Research. 1994;**22**(11):2065-2070

[38] Rybchin VN, Svarchevsky AN. The plasmid prophage N15: A linear DNA with covalently closed ends. Molecular Microbiology. 1999;**33**:895-903

[39] Landschulz WH, Johnson PF, Mcknight SL, Landschulz WH, Johnson PF, Mcknight SL. Structure common to a new class of the leucine zipper: A hypothetical DNA binding proteins. Science. 1988;**240**(4860):1759-1764

[40] Ravin NV, Strakhova TS, Kuprianov V. The protelomerase of the phage-plasmid N15 is responsible for its maintenance in linear form. Journal of Molecular Biology. 2001;**312**(5):899-906

[41] Feiss M, Young J, Sultana S, Patel P, Sippy JDNA. Packaging specificity of bacteriophage N15 with an excursion into the genetics of a cohesive end mismatch. PLoS ONE. 2015;**10**(12):e0141934

[42] Papagiannis CV, Sam MD, Abbani MA, Yoo D, Cascio D, Clubb RT, et al. Fis targets assembly of the Xis nucleoprotein filament to promote excisive recombination by phage lambda. Journal of Molecular Biology. 2007;**367**(2):328-343

[43] Replication RNV. Maintenance of linear phage-plasmid N15. Microbiology Spectrum. 2015;**3**(1):1-12

[44] Baril C, Richaud C, Baranton G. Linear chromosome of *Borrelia burgdorferi*. Research in Microbiology. 1989;**140**(8):507-516

[45] Picardeau M, Lobry JR, Joseph B, Bernard C. Physical mapping of an origin of bidirectional replication at the centre of the *Borrelia burgdorferi* linear chromosome. Molecular Microbiology. 1999;**32**(2):437-445

[46] Chaconas G, Stewart PE, Tilly K, Bono JL, Rosa P. Telomere resolution in the Lyme disease spirochete. The EMBO Journal. 2001;**20**(12):3229-3237

[47] Byram R, Stewart PE, Rosa P. The essential nature of the ubiquitous 26-kilobase circular replicon of *Borrelia burgdorferi*. Journal of Bacteriology. 2004;**186**(11):3561-3569

[48] Yang W, Mizuuchi K. Site-specific recombination in plane view. Structure. 1997;**5**(11):1401-1406

[49] Cheng C, Kussie P, Pavletich N, Shuman S. Conservation of structure and mechanism between eukaryotic topoisomerase I and site-specific recombinases. Cell. 1998;**92**:841-850

[50] Krogh BO, Shuman S. A poxvirus-like type IB topoisomerase family in bacteria. Proceedings of the National Academy of Sciences. 2002;**99**(4):1853-1858

[51] Kobryn K, Briffotaux J, Karpov V. Holliday junction formation by the *Borrelia burgdorferi* telomere resolvase, ResT: Implications for the origin of genome linearity. Molecular Microbiology. 2009;**71**(5):1117-1130

[52] Grainge I, Jayaram M. The integrase family of recombinases: Organization and function of the active site. Molecular Microbiology. 1999;**33**:449-456

[53] Esposito D, Scocca JJ. The integrase family of tyrosine recombinases: Evolution of a conserved active site domain. Nucelic Acids Research. 1997;**25**(18):3605-3614

[54] Tourand Y, Lee L, Chaconas G. Telomere resolution by *Borrelia burgdorferi* ResT through the collaborative efforts of tethered DNA binding domains. Molecular Microbiology. 2007;**64**:580-590

[55] Lee J, Tonozuka T, Jayaram M. Mechanism of active site exclusion in a site-specific recombinase: Role of the DNA substrate in conferring half-of-the-sites activity. Genes & Development. 1997;**11**:3061-3071

[56] Nash HA, Robertson CA. Heterduplex substrates for bacteriophage lambda site-specific recombination: Cleavage and strand transfer products. The EMBO Journal. 1989;**8**(11):3523-3533

[57] Deneke J, Burgin AB, Wilson SL, Chaconas G. Catalytic residues of the telomere resolvase ResT. The Journal of Biological Chemistry. 2004;**279**(51):53699-53706

[58] Chen Y, Rice PA. The role of the conserved Trp 330 in Flp-mediated recombination. The Journal of Biological Chemistry. 2003;**278**(27):24800-24807

[59] Nunes-düby SE, Kwon HJ, Tirumalai RS, Ellenberger T, Landy A. Similarities and differences among 105 members of the Int family of site-specific recombinases. Nucleic Acids Research. 1998;**26**(2):391-406

[60] Wang JC. DNA topoisomerases. Annual Review of Biochemistry. 1996;**65**:635-692

[61] Kormann MSD, Hasenpusch G, Aneja MK, Nica G, Flemmer AW, Herber-jonat S, et al. Expression of therapeutic proteins after delivery of chemically modified mRNA in mice. Nature Biotechnology. 2011;**29**(2):154-157

[62] Bangel-ruland N, Fernández EF, Leier G, Leciejewski B, Rudolph C, Rosenecker J, et al. Cystic fibrosis transmembrane conductance regulator-mRNA delivery: A novel alternative for cystic fibrosis gene therapy. The Journal of Gene Medicine. 2013;**15**:414-426

[63] Gallie D. The cap and poly (A) tail function synergistically to regulate mRNA translational efficiency. Genes & Development. 1991;**5**:2108-2116

[64] Grier AE, Burleigh S, Sahni J, Clough CA, Cardot V, Choe DC, et al. pEVL: A linear plasmid for generating mRNA IVT templates with extended encoded poly(A) sequences. Molecular Therapy--Nucleic Acids. 2016;**5**(4):e306

[65] Wong S, Lam P, Nafissi N, Denniss S, Slavcev R. Production of double-stranded DNA ministrings. Journal of Visualized Experiments. 2016;**108**:3-9

[66] Catanese DJ, Fogg JM, Schrock DE, Gilbert BE, Zechiedrich L. Supercoiled minivector DNA resists shear forces associated with gene therapy delivery. Gene Therapy. 2011;**19**(1):94-100

[67] Liang X, Baek C, Katzen F. *Escherichia coli* with two linear chromosomes. ACS Synthetic Biology. 2013;**2**(12):734-740

[68] Scott VL, Patel A, Villarreal DO, Hensley SE, Ragwan E, Yan J, et al. Novel synthetic plasmid and doggybone DNA vaccines induce neutralizing antibodies and provide protection from lethal influenza challenge in mice. Human Vaccines & Immunotherapeutics. 2015;**11**(8):1972-1982

[69] Allen A, Wang C, Caproni LJ, Sugiyarto G, Harden E, Douglas LR, et al. Linear doggybone DNA vaccine induces similar immunological responses to conventional plasmid DNA independently of immune recognition by TLR9 in a pre-clinical model. Cancer Immunology, Immunotherapy. 2018;**67**(4):627-638

[70] Daya S, Berns KI. Gene therapy using adeno-associated virus vectors. Clinical Microbiology Reviews. 2008;**21**(4):583-593

[71] Naso MF, Tomkowicz B, Perry WL, Strogl WR. Adeno-associated virus (AAV) as a vector for gene therapy. BioDrugs. 2017;**31**(4):315-332

[72] Hammerl JA, Klein I, Appel B, Hertwig S. Interplay between the temperate phages PY54 and N15, linear plasmid prophages with covalently closed ends. Journal of Bacteriology. 2007;**189**(22):8366-8370

Chapter 5

Scale-Up and Bioprocessing of Phages

John Maxim Ward, Steven Branston, Emma Stanley and Eli Keshavarz-Moore

Abstract

A profusion of new applications for phage technologies has been developed within the last few years, stimulating investigations into the large-scale production of different phages. Applications such as antibiotic replacement, phages as gene therapy vectors, phages as vaccines, diagnostics using filamentous phages and novel optical applications such as the phage laser may need grams to kilogrammes of phage in the future. However, many of the techniques that are used for the growth and purification of bacteriophage at small scale are not transferable to large-scale production facilities of phage in industrial processes. In this chapter, the stages of production that need to be carried out at scale are examined for the efficient large-scale fermentation of the filamentous phage M13 and the *Siphoviridae* phage lambda (λ). A number of parameters are discussed: the multiplicity of infection (MOI) of phage to host cells, the impact of agitation on the initial infection stages, the co-growth with phage rather than static attachment, the use of engineered host cells expressing nuclease, the optimisation of both the quantity and the physiology of the *E. coli* inoculum and phage precipitation methods.

Keywords: phage, PEG precipitation, nuclease, filamentous phage, lambda, phage diagnostics, phage laser, fermenter

1. Introduction

Bacteriophages, often shortened to just phages, are viruses that infect bacteria. Their discovery and characterisation in the early days of bacterial molecular biology has led to certain phages being very well understood in terms of their life cycle, and several phages that infect *Escherichia coli* have become tools in molecular biology techniques such as cloning [1–3]. There has been a resurgence recently in the use of bacteriophages as therapeutics, as vectors for the delivery of vaccines [4], for the killing of pathogenic bacteria as an alternative to antibiotics [5] and for gene therapy to transfer DNA to target human or animal cells [6]. Some of these uses would need the production of many millions of doses of a vaccine, for example, or very large quantities for use as an antibacterial. This has increased demand for investigation into the large-scale production of bacteriophage which would necessitate volumes from hundreds to thousands of litres. The use of phage as biotherapeutics such as vaccines or for gene therapy may be advantageous as phage is considered cheap to manufacture, with large quantities of the product being rapidly produced. But the large-scale production of wild type or genetically modified bacteriophages for use in the biotherapeutics industry provides significant process and regulatory challenges. Bacteriophages, like any virus, are dependent on a host organism to propagate, in the examples here it is *E. coli*; consequently, the generation of progeny bacteriophage is unequivocally linked to the physiology, molecular biology and growth needs of the host which are important to understand in order to maximise production.

Methods for the production of phages, e.g. λ and M13 bacteriophage, at laboratory production scale have remained unchanged for many years [7]. However, aspects of these protocols are either not practical or unsuitable for large-scale production of phages. Therefore, it is highly desirable to consider early on in the development of phage technologies how any successful bacteriophage therapeutic would be produced at large scale at an industrial level.

One of the problems associated with producing and using λ as a biotherapeutic is the issue of host-derived nucleic acid. The λ lifecycle [8] involves the phage progeny escaping from the host cell by lysis of the bacteriophage host, whereupon the cell contents including high-molecular-weight host chromosomal DNA and RNA are released into the culture supernatant. This significant quantity of host cell-derived nucleic acid can cause important problems for both downstream processing [9] and from a regulatory point of view [10], so reducing the presence of bacterial host nucleic acid in the first stages of the process stream would remove these issues.

M13 is an unusual phage because it does not lyse its host and the entire phage is secreted from the host bacterium through special pores spanning the cell wall [11] although this does make the culture supernatant relatively free of contaminating host cell material, unlike the supernatant in a λ fermentation. In both lytic and secreted phage production the first downstream stage is the concentration of the phage from whatever volume of growth medium was used to grow the infected cells. Filamentous phage such as M13 has a very asymmetric shape with wild-type M13 having a length to width ratio of 138:1 and this extreme asymmetry allows a mild precipitation using polyethylene glycol (PEG) [12].

In this paper we present initial studies into the parameters that need to be manipulated for scaling up fermentation of M13 phage for industrial production.

2. Lambda phage

Lambda (λ) bacteriophage is a temperate phage with a double-stranded (ds) genome of approximately 48 kb [13]. This is encapsulated in an icosahedral capsid (~50 nm in diameter) with a long flexible non-contractile tail (~150 nm in length). The host for λ production is *E. coli* with infection by lambda phage taking place via the maltose binding protein, LamB. Lambda bacteriophage is one of the most intensely studied bacteriophages and has been used for many studies on uncovering basic molecular biology [14] and in biotechnology for phage display of peptides and proteins [15], vaccine [16] and gene transfer and therapy [6].

The large-scale production of genetically modified lambda bacteriophage for use in the biotherapeutics industry provides significant process and regulatory challenges.

One of the problems associated with producing and using lambda bacteriophage as a biotherapeutic is the issue of host-derived nucleic acid. The lambda lifecycle involves the cell lysis of the bacteriophage host, whereupon high-molecular-weight host chromosomal DNA is liberated into the culture supernatant. The presence of large quantities of host cell-derived nucleic acid can cause significant problems during processing as high-molecular-weight chromosomal DNA causes an increase in the cell lysate viscosity [9]. Furthermore, the presence of nucleic acid in the final product is non-desirable from a regulatory perspective [10], and thus reducing its presence in the first stages of the process stream would alleviate these issues. The lysis of the host *E. coli* cell and the release of intracellular contents (DNA, RNA and proteins) as well as fragments of cell wall will have detrimental effects when processing lambda. For example, intracellular contents can co-precipitate with the phage, can compete for binding sites on chromatography material and can block membranes and chromatography columns. These contaminants need to be taken into consideration when planning a large-scale purification and downstream processing protocol.

3. M13: a filamentous bacteriophage

M13 is an unusual phage as it has a filamentous structure of 900 nm in length and 6.5 nm in width. It is a member of a small group of closely related phages including F1 and Fd [17] that infect *E. coli*. The genome is a single-stranded circular DNA molecule, and the length of the phage (but not its width) is simply determined by the size of the page genome. Short phage particles can be made using plasmids that contain just the replication origin and packaging signals, and phage particles longer than the wild type can be made by inserting DNA sequences into the phage genome. It was thought that the very long but thin shape of M13 and other filamentous phages would increase their shear sensitivity in the various kinds of industrial-scale processing equipment of pumps, continuous centrifuges and membrane filters. This was seen not to be the case [18] which is highly advantageous for large-scale downstream processing of this and other filamentous phages. Filamentous phages have a rather special property in that they do not lyse their host, but set up a permanently infected state, and new, progeny phage is extruded through special structures in the cell wall. Derivatives of M13 phage were extensively used in the early years of DNA sequencing by the Sanger method [19] and in the techniques of site-directed mutagenesis [20] and phage display [21] for the maturation of recombinant antibodies [22].

The unusual mode of growth of filamentous phage by secretion from the host without lysis has considerable advantages for these molecular biology methods because the phage in the supernatant of growing cells is relatively free of any cellular contaminants [23] such as intracellular proteins, genomic DNA and RNA. This makes the purification of filamentous phage a relatively simple matter with many fewer contaminants than phage λ cultures.

4. Multiplicity of infection (MOI)

The multiplicity of infection or MOI is the number of phage particles added per host cell to initiate infection and thus production. The methods developed for the uses of phage λ and M13 at small scale or a few mL generally tend to use a high MOI of 5–10 or more. The MOI is important for scale-up as it defines how large the culture that provides the phage for the scale-up needs to be. It is neither desirable nor sensible to have to grow a fermenter of phage to inoculate a slightly larger fermenter in the final preparation. Also with a large MOI of just under 10, we reach the point where there will be enough phage for every cell to be infected, and at that MOI we can only expect a single burst of phage particles for lytic phage and therefore only a small increase in the number of phage added. With an MOI of 1, only 63% of cells will be infected by one or more phages because there is a Poisson distribution of MOI to infected cells [24]. But some phages have many binding sites per cell, e.g.

T4 has 10^5 molecules of OmpC to bind to [25], while M13 only has approximately 3 [26], so the kinetics of phage finding and attaching to bacteria and forming a productive infection are quite complex. There are 30,000 LamB proteins in the outer membrane which is the initial receptor for λ but many fewer copies of the mannose receptor, ManY, in the inner membrane which is where the DNA of λ crosses the inner membrane [27, 28]. For M13 the receptor is the tip of the F plasmid's transfer pilus, and there are usually one or two F pili per cell. But once M13 has established its replication inside an *E. coli* host, the cell is then permanently infected and will continue to secrete phage from these intact cells.

Under ideal conditions the burst size of λ phage particles is 170 +/− 10 which takes 51 min [29], and during this time uninfected cells will still be growing and dividing, providing new hosts for the phage that are released. To get repeated rounds of replication, the ideal cell numbers and MOI for each phage are different and take into account cell division rates, numbers of receptors, the choice between a lytic and lysogenic cycle in phages where those can take place, the burst size and rate of replication and maturation of the phage. This complex interaction of several parameters means that it is difficult to say a priori what the combination of cells, phage and time of addition is the most appropriate for a given size of growth chamber.

Figure 1 shows the relationship between the host *E. coli* and MOI of M13.

The graph of MOI and final phage production in **Figure 1** shows that from a 10^6 range of MOI added at the start of growth in **Figure 1**, all three cultures reach the same final M13 phage titre. It just takes slightly longer for the lower MOI cultures to reach the final of approx. 5×10^{11} phage per mL. This has important consequences for scale-up of M13 production. For example, if a large fermenter of, say, 100 L were needed and the MOI of 50:1 was needed, we would need to have 500 mL of the equivalent inoculum used here. The information from **Figure 1** shows that we can use just 0.5 mL of the same titre inoculum or much less, e.g. down to a few microlitres. For convenience it is best to inoculate a fermenter with enough in terms of volume that will reach the medium in the fermenter, so a few millilitres of a phage dilution are all that is needed. This means that one phage stock can be used for multiple fermentation runs.

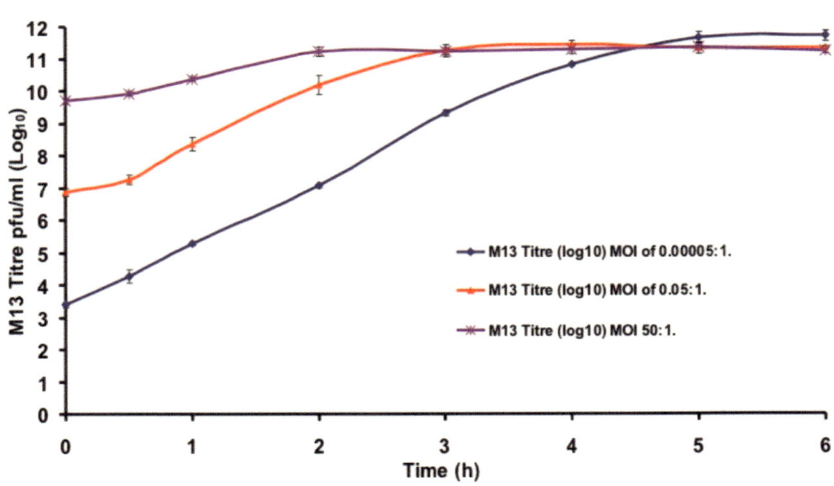

Figure 1.
Titre of M13 produced from different MOI on E. coli TOP10 F'. M13 phage was prepared from a 400 mL culture of E. coli TOP10 F' and precipitated with PEG 6000 at a final concentration of 3.3% and 330 mM NaCl. The precipitated phage was centrifuged at 14,000 ×g for 10 min at 4°C and resuspended in 8 mL of 10 mM Tris.HCl pH 7.5 and filtered through a 0.22 μm filter. The stock of M13 was approximately 1 × 10^{12} pfu/mL. E. coli TOP10 F' was grown in 400 mL of Nutrient Broth No. 2 (Oxoid) and a 40-mL inoculum in 2-L shake flask at 37°C. Around 2 mL of M13 was added at the appropriate dilutions to achieve the three different MOI of 0.00005:1, 0.05:1 and 50:1. Each point is the average of three flasks.

5. Initial phage binding and infection in fermenters

The standard way of initiating infection of λ or M13 is to mix the phage and cells and allow a static incubation for usually 5–15 min for phage attachment to the phage binding target on the cell surface. The phage/cell mixture is then added to broth if liquid growth is desired or to 3 mL of soft agar and poured onto an agar plate if plaques are wanted. This static incubation is in the recipes for all phages being handled at the lab scale and probably came about because researchers thought it would maximise the attachment of phage to their host. However, in a fermenter it is not sensible to turn off the stirrer and let the cells sit for 15 min while phage attachment takes place. We tested what would happen if M13 phage were simply added directly to shaking cultures of *E. coli* JM107 without stopping the shaking and comparing this with a static attachment/incubation of 30 min.

Figure 2 shows this experiment with three different inocula of M13 phage giving MOI ranging over a 10^6 range.

It is clear from **Figure 2** that the static incubation is not needed for the initial attachment of M13 to sensitive *E. coli*, and so the required dose of M13 phage can be directly added to a fermenter with the *E. coli* host already growing with the impeller stirring. The culture which will be stirred at a high rate *does not* need to be stopped and left to go static for 15–30 min. This would compromise the growth of the cells in a large fermenter, and so our findings give a positive help for the scale-up of phage and the way one can run a large fermenter where procedures have to be different from what is done at small scale in the molecular biology lab.

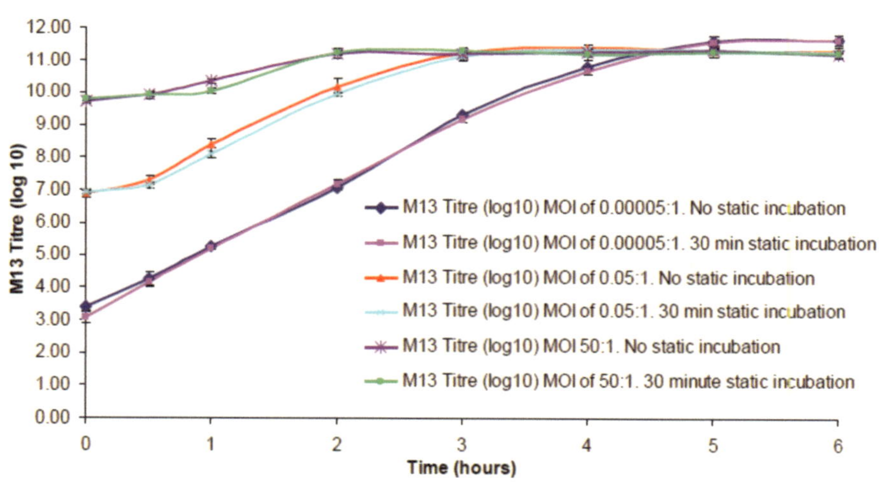

Figure 2.
*Static versus shaking incubation of M13 with E. coli TOP10 F'. The conditions and culture quantities were identical to those shown in **Figure 1** with the addition of three conditions at each MOI where the M13 phage was added and left for 30 min static incubation. These cultures for the static incubation were then grown with shaking at 200 rpm in a 37°C orbital incubator. Each point is the average of three flasks.*

6. Nuclease-producing *E. coli*

The above sections on MOI and removing the necessity for a static attachment of phage were shown with M13 as examples. The M13 phage does not lyse its host, so the culture medium after infection is uncontaminated by the bulk of the host cellular contents and largely contains just the filamentous M13 bacteriophage particles. We have examined the supernatants from M13-infected cultures and determined the levels of host cell contaminants of DNA and protein [23] which are quite low compared to the large amount of cellular DNA, RNA and protein released by lytic phage. A phage such as λ is a representative of lytic phage which is the majority of the types of phages used in therapy and biological control for the replacement of antibiotics. At each cycle of replication and release of the new phage particles, the host is lysed, and the cell contents are released into the medium along with insoluble debris from the cell wall and membrane. This leads to problems in subsequent downstream purification due to the large amount of different cellular molecules competing in the subsequent downstream processing steps.

The host RNA and DNA represent major contaminants that need to be removed especially for gene therapy applications. The release of host cell DNA also increases the viscosity of the medium, and this has an adverse effect on clarification by centrifugation and membrane concentration due to blocking of the membranes. For lab-scale molecular biology work, it is normal to add the enzymes pancreatic RNAseI and pancreatic DNAseI from bovine pancreas preparations. With the advent of bovine spongiform encephalopathy (BSE) which peaked in the 1990s, the addition of any bovine or animal proteins into the growth or purification train of material destined for human therapy was banned. These regulatory restrictions removed the ability to use these cheap nucleic acid-degrading enzymes, and the substitutes from bacterial sources were much more expensive. A strategy to overcome this problem was developed by us, and this was to express a broad-spectrum nuclease in the periplasm of *E. coli* which would be released as cells were lysed [30, 31]. The enzyme Staphylococcal nuclease has been extensively characterised and used from the 1960s onwards [32] and can degrade both DNA and RNA. The expression of this in *E. coli* where it is secreted into the periplasm does not affect the growth of *E. coli* because the enzyme cannot get access to its substrate while the cell is growing normally. When the cells are lysed by a bacteriophage such as λ or by homogenisation, the nuclease can gain access to the released DNA and RNA and degrade them. This 'cell engineering' approach to assisting bioprocessing was developed at UCL and has been shown to give considerable gains in the centrifugation steps and other downstream purification steps in bioprocessing of proteins such as Fab fragments [33–35].

We sought to apply this cell engineering strategy for the production of λ phage and to help solve the problem of the large amount of cellular contaminants released into the media when λ phage cultures need to be harvested and processed.

E. coli JM107 [3] was grown in 2 L fermenters with either JM107 or JM107 containing the plasmid pMMBompnucB which is a broad host-range plasmid vector based on an Inc. Q plasmid RSF1010 and contains the Staphylococcal nuclease which has been altered by the addition of the *E. coli* ompA signal sequence for secretion [31]. **Figure 3** shows the growth of the two hosts with no added λ phage and the same hosts with 8×10^{10} λ phage particles added after 2 hours when the OD_{600} had reached 10.

The addition of λ to the fermenter used the strategy that we had developed where a low MOI is used and without a static initial incubation of the host cells and the phage. In this way we can add the phage directly to a growing fermenter of host with the impeller (single shaft with three top-driven, equally spaced, six-bladed turbines) and four diametrically opposed baffles in the fermenter, running throughout. The growth profiles of the two uninfected cultures show no difference in their growth profiles which means that the presence of the expressed nuclease enzyme in the periplasm has no effect on growth rate or final OD. In the two cultures with added λ phage, the OD drop is the same for both hosts showing that λ replication and cell lysis are the same in both.

The production of λ was monitored throughout growth and is shown in **Figure 4**.

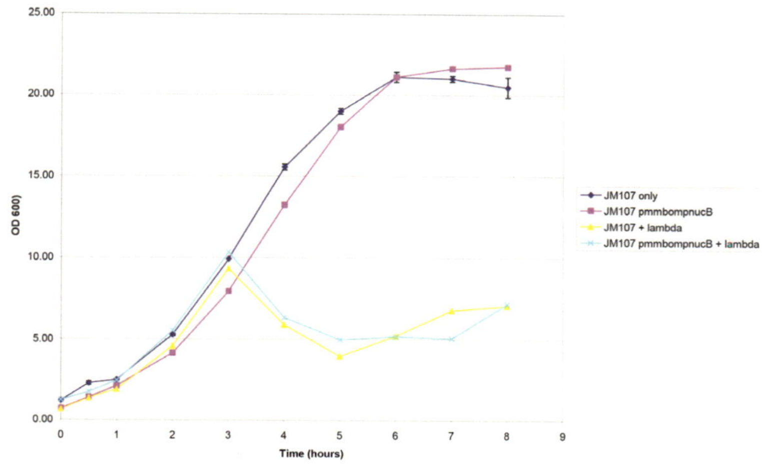

Figure 3.
Growth of E. coli JM107 and λ production with and without a nuclease-expressing plasmid. E. coli JM107 containing pMMBompnucB and with no plasmid was grown in 2 L (working volume 1.5 L) LH 210 series fermenter (Bioprocess Engineering Services, Charing, Kent, UK) with 150 mL of E. coli inoculum and a final working volume of 1.5 L in phage media containing 100 μg/mL ampicillin and 20 μg/mL IPTG. λ phage particles were added at 2 hours to give an MOI of 0.05 (4 mL of 2 × 10^{10} pfu/mL). Impeller stirring continued throughout the growth and addition of phage, and the OD_{600} was monitored.

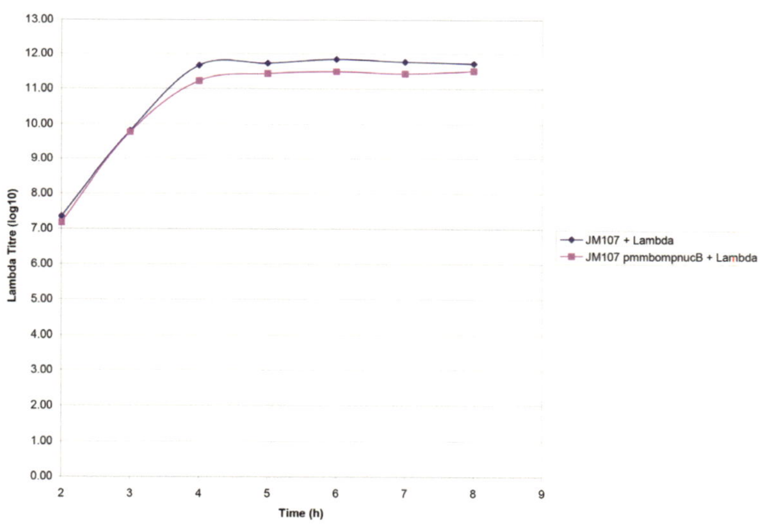

Figure 4.
*Production of λ phage in 2 L fermenters with and without periplasmic nuclease expression. λ phage was added at 2 hours in **Figure 3**, and the graph here shows λ phage amounts from the 2 hours onwards. Samples from the fermenters shown in **Figure 3** were diluted in phage buffer and titred on E. coli JM107 using soft top agar and plaque counting.*

Both the nuclease and the non-nuclease-expressing *E. coli* JM107 produced the same 5-log increase in phage particles with the same time profile. The efficacy of the nuclease in the removal of the host nucleic acid was assessed by the electrophoresis of samples from each time point on agarose gels and the visualisation of the released nucleic acid (both DNA and RNA). **Figure 5** shows the complete degradation and removal of the released host genomic DNA and host RNA in the strain of *E. coli* expressing the periplasmic nuclease.

The presence of the expressed periplasmic nuclease is apparent from the difference in the samples in **Figure 5B** compared to the samples from the same time points in the fermenter with no expression plasmid for the Staphylococcal nuclease.

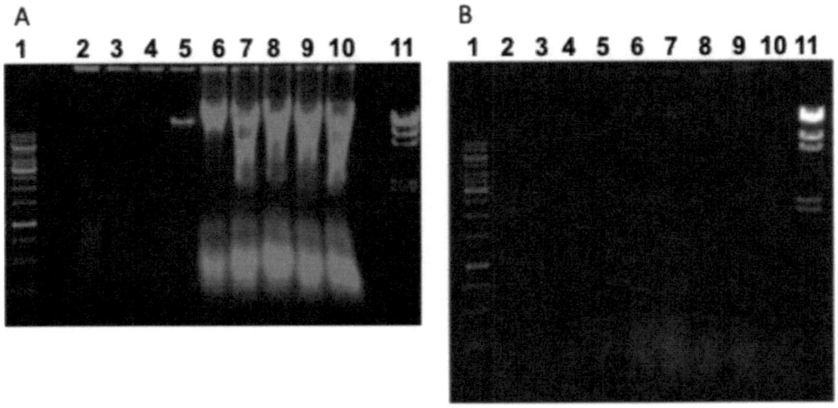

Figure 5.
Degradation of host nucleic acid from l infected E. coli JM107 and JM107 containing pMMBompnucB. (A) Agarose gel with samples from fermenter in ***Figure 3*** *growing E. coli JM107 with λ. 1, 1 kb ladder; 2, 0 h; 3, 1 h; 4, 2 h; 5, 3 h; 6, 4 h; 7, 5 h; 8, 6 h; 9, 7 h; 10, 8 h; 11 λ HindIII ladder. (B) Agarose gel with samples from fermenter in* ***Figure 3*** *growing E. coli JM107 containing pMMBompnucB infected with λ. 1, 1 kb ladder; 2, 0 h; 3, 1 h; 4, 2 h; 5, 3 h; 6, 4 h; 7, 5 h; 8, 6 h; 9, 7 h; 10, 8 h; 11 λ HindIII ladder. The same volume of sample from each fermenter time point was loaded onto each lane.*

Almost all of the released host genomic DNA and the large majority of the RNA has been degraded in the culture that expresses the nuclease. **Figure 4** shows that the production of λ phage particles is identical in both fermenters, and the presence of the nuclease does not impinge on λ production and leads to a removal of the majority of the nucleic acid that would normally need the addition of bovine pancreatic DNAseI and RNAseA or more expensive bacterial equivalents such as Benzonase™ [36] to reduce the amount of nucleic acids. This cell engineering approach means that no animal-derived enzymes need be added, no costly commercial bacterial enzymes need be added and the engineered cells provide their own nuclease which degrades the nucleic acid in situ, so no additional time for incubation of any added enzyme is needed. Therefore a saving of both time and money is achieved via cell engineering for bioprocessing.

7. Precipitation of M13

Bacteriophages produced at any scale need to be concentrated by some method after the growth and production have taken place. The properties of phages allow some precipitation methods that are milder than conditions needed to precipitate host soluble protein or nucleic acid. Phages are large multicomponent entities usually many hundreds of times larger than a medium-sized soluble protein. Their asymmetric shape also enables mild precipitation methods. Polyethylene glycol (PEG) precipitation is a mild method of precipitating biological material and is very efficacious in the precipitation of large asymmetric material such as DNA or macromolecular assemblies, e.g. virus-like particles such as phages [37, 38]. The larger and more asymmetric the particle, the lower the amount of PEG is needed to precipitate the particle and leave behind other smaller soluble materials. An exploration of different average molecular weights of PEG from 600 to 20,000 for M13 precipitation showed that PEG 6000 and PEG 8000 combined the best properties of precipitation at low % concentration with lower viscosities than PEG 12,000 and PEG 20,000 [23] and 2% PEG 6000 with 330 mM NaCl gave >95% precipitation of M13. The relationship between PEG and NaCl is shown in **Figure 6** where the increasing PEG molecular weight and PEG concentration with % of M13 recovered were investigated.

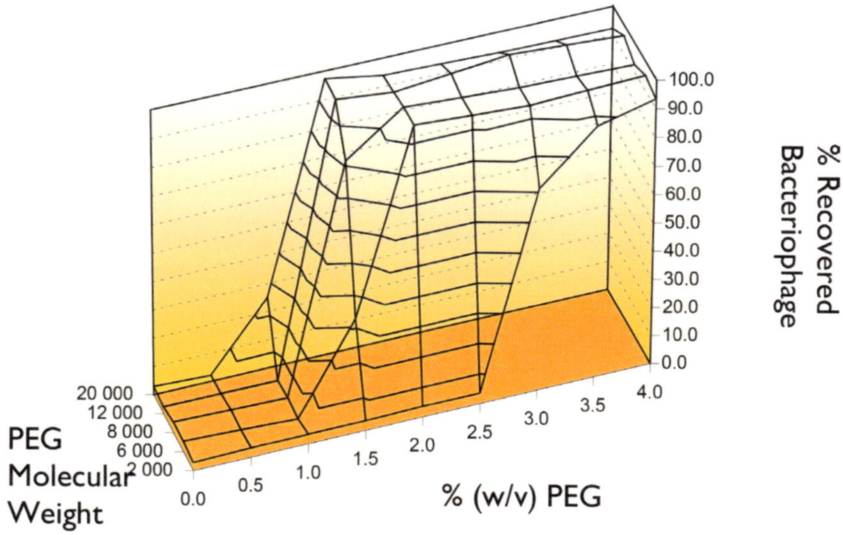

Figure 6.
The relationship between M13 recovery and PEG molecular weight and percentage PEG. All precipitations were carried out at 10 mL scale with additions of PEG to the final % concentration shown in the diagram with 4 × 10¹² CsCl purified M13 and 330 mM NaCl. The mixtures were incubated on ice for 1 hour and centrifuged for 10 min at 8000 ×g at 4°C.

The diagram in **Figure 6** shows the relationship between increasing chain length/molecular weight of PEG, the % of PEG and the amount of M13 phage recovered. Higher-molecular-weight PEG preparations can be used at low final percentages, but there is significant increase in the viscosity of high-molecular-weight PEG solutions in the stock solutions needed to add to the phage-containing solution, making PEG 20,000 very difficult to use. PEG 6000 and PEG 8000 achieve virtually the same precipitation profile as PEG 12,000 and the lower viscosity of stock solutions of PEG 6000 make this chain length the best for precipitation with low viscosity. A concentration of 2% w/v PEG 6000 is ideal with 330 mM NaCl [23]. It was discovered that the nutrient media commonly used for growth such as NB2 contains sufficient salt (Na, K and NH_3 ions) that the added NaCl can be reduced to only 135 mM which would make a saving in materials and disposal costs in large-scale M13 phage precipitation.

8. Conclusion

The strategies for the scale-up of phage growth and primary downstream purification are still in their infancy, but we have shown that there are significant gains to be made from the work described here. The amount of phage that needs to be added to large-scale growth volumes can be reduced by several orders of magnitude from what is common at the lab scale. Phage can be added directly to fermenters where rapidly stirring impellers are needed to maintain aeration and correct physiology of the host with no cessation of the stirring, and the phage will find their host and attach with no difficulty. The use of cells engineered to produce their own broad-spectrum periplasmic nuclease gives significant gains in the destruction of host DNA and RNA release on lysis, and this prevents the contamination of the phage with nucleic acid when the phages are concentrated by precipitation. The gentle precipitation method of using low concentrations of PEG can then be used to give relatively pure preparations of phage in one step. These methods can be used together and will allow the large-scale uses of phage in the future in medical and clinical applications and then beyond into biotechnological applications such as the uses of filamentous phage in electronics like phage batteries [39] and the phage laser [40].

Acknowledgements

We acknowledge and thank the Engineering and Physical Sciences Research Council (EPSRC) for support via the Life Science IMRC for Bioprocessing and the EPSRC for the PhD studentship to SB. We thank the Biotechnology and Biological Sciences Research Council (BBSRC) for grant funding to JMW and EKM to support ES under research grant BBD521465/1.

Conflict of interest

There are no conflicts of interest.

Author details

John Maxim Ward*, Steven Branston, Emma Stanley and Eli Keshavarz-Moore
Department of Biochemical Engineering, University College London, London, UK

*Address all correspondence to: j.ward@ucl.ac.uk

References

[1] Karn J, Brenner S, Barnett L, Cesareni G. Novel bacteriophage lambda cloning vector. Proceedings of the National Academy of Sciences of the United States of America. 1980;**77**:5172-5176

[2] Chauthaiwale VM, Therwath A, Deshpande VV. Bacteriophage lambda as a cloning vector. Microbiological Reviews. 1992;**56**:577-591

[3] Yanisch-Perron C, Vieira J, Messing J. Improved M13 phage cloning vectors and host strains: Nucleotide sequences of the M13mp18 and pUC19 vectors. Gene. 1985;**33**:103-119

[4] Lankes HA, Zanghi CN, Santos K, Capella C, Duke CM, Dewhurst S. *In vivo* gene delivery and expression by bacteriophage lambda vectors. Journal of Applied Microbiology. 2007;**102**:1337-1349

[5] Azeredo J, Sutherland IW. The use of phages for the removal of infectious biofilms. Current Pharmaceutical Biotechnology. 2008;**9**:261-266

[6] Jepson CD, March JB. Bacteriophage lambda is a highly stable DNA vaccine delivery vehicle. Vaccine. 2004;**22**:2413-2419

[7] Green MR, Sambrook J. Molecular Cloning. New York: Cold Spring Harbor Laboratory Press; 2012

[8] Casjens SR, Hendrix RW. Bacteriophage lambda: Early pioneer and still relevant. Virology. 2015;**479-480**:310-330

[9] Boynton ZL, Koon JJ, Brennan EM, Clouart JD, Horowitz DM, Gerngross TU, et al. Reduction of cell lysate viscosity during processing of poly (3-hydroxyalkanoates) by chromosomal integration of the Staphylococcal nuclease gene in *Pseudomonas putida*. Applied and Environmental Microbiology. 1999;**65**:1524-1529

[10] World Health Organisation. Guidelines on the quality, safety, and efficacy of biotherapeutic protein products prepared by recombinant DNA technology. World Health Organisation. Replacement of Annex 3 of WHO Technical Report Series, No. 814; 2014

[11] Smeal SW, Schmitt MA, Pereira RR, Prasad A, Fisk JD. Simulation of the M13 life cycle I: Assembly of a genetically-structured deterministic chemical kinetic simulation. Virology. 2017;**500**:259-274

[12] Du Z, Hood L, Wilson RK. Automated fluorescent DNA sequencing of polymerase chain reaction products. Methods in Enzymology. 1993;**218**:104-121

[13] Sanger F, Coulson AR, Hong GF, Hill DF, Petersen GB. Nucleotide sequence of bacteriophage λ DNA. Journal of Molecular Biology. 1982;**162**:729-773

[14] Gottesman ME, Weisberg RA. Little Lambda, who made thee? Microbiology and Molecular Biology Reviews. 2004;**68**:796-813

[15] Maruyama IN1, Maruyama HI, Brenner S. Lambda foo: A lambda phage vector for the expression of foreign proteins. Proceedings of the National Academy of Sciences of the United States of America 1994; **91**:8273-8277

[16] Razazan A, Nicastro J, Slavcev R, Barati N, Arab A, Mosaffa F, et al. Lambda bacteriophage nanoparticles displaying GP2, a HER2/neu derived peptide, induce prophylactic and therapeutic activities against TUBO tumor model in mice. Scientific Reports. 2019;**9**:2221

[17] Mai-Prochnow A, Gee J, Hui K, Kjelleberg S, Rakonjac J, McDougald D, et al. Big things in small packages: The genetics of filamentous phage and effects on fitness of their host. FEMS Microbiology Reviews. 2015;**39**:465-487

[18] Branston S, Stanley E, Ward J, Keshavarz-Moore E. Study of robustness of filamentous bacteriophages for industrial applications. Biotechnology and Bioengineering. 2011;**108**:1468-1472

[19] Gardner RC, Howarth AJ, Messing J, Shepherd RJ. Cloning and sequencing of restriction fragments generated by Eco RI*. DNA. 1982;**1**:109-115

[20] Norrander J, Kempe T, Messing J. Construction of improved M13 vectors using oligodeoxynucleotide-directed mutagenesis. Gene. 1983;**26**:101-106

[21] Smith GP. Filamentous fusion phage: Novel expression vectors that display cloned antigens on the virion surface. Science. 1985;**228**:1315-1317

[22] Clackson T, Hoogenboom HR, Griffiths AD, Winter G. Making antibody fragments using phage display libraries. Nature. 1991;**352**:624-628

[23] Branston S, Stanley E, Keshavarz-Moore E, Ward J. Precipitation of filamentous bacteriophages for their selective recovery in primary purification. Biotechnology Progress. 2011;**28**:129-136

[24] Ellis EL, Delbruck MJ. The growth of bacteriophage. The Journal of General Physiology. 1939;**22**:365-384

[25] Xu Z, Lee SY. Display of polyhistidine peptides on the *Escherichia coli* cell surface by using outer membrane protein C as an anchoring motif. Applied and Environmental Microbiology. 1999;**65**:5142-5147

[26] Clarke M, Maddera L, Harris RL, Silverman PM. F-pili dynamics by live-cell imaging. PNAS. 2008;**105**:17978-17981

[27] Gibbs K, Isaac D, Xu J, Hendrix R, Silhavy T, Theriot J. Complex spatial distribution and dynamics of an abundant *Escherichia coli* outer membrane protein, LamB. Molecular Microbiology. 2004;**53**:1771-1783

[28] Edgar R, Rokney A, Feeney M, Semsey S, Kessel M, Goldberg MB, et al. Bacteriophage infection is targeted to cellular poles. Molecular Microbiology. 2008;**68**:1107-1116

[29] Wang I-N. Lysis timing and bacteriophage fitness. Genetics. 2006;**172**:17-26

[30] Cooke GD, Cranenburgh RM, Hanak JAJ, Dunnill P, Thatcher DR, Ward JM. Purification of essentially RNA free plasmid DNA using a modified *Escherichia coli* host strain expressing Ribonuclease A. Journal of Biotechnology. 2001;**85**:297304

[31] Cooke GD, Cranenburgh RM, Hanak JAJ, Ward JM. A modified *Escherichia coli* protein production strain expressing staphylococcal nuclease, capable of auto-hydrolysing host nucleic acid. Journal of Biotechnology. 2003;**101**:229-239

[32] Heins JN, Suriano JR, Taniuchi H, Anfinsen CB. Characterization of a nuclease produced by *Staphylococcus aureus*. The Journal of Biological Chemistry. 1967;**242**:1016-1020

[33] Balasundaram B, Nesbeth D, Ward JM, Keshavarz-Moore E, Bracewell DG. Step change in the efficiency of centrifugation through cell engineering: Coexpression of Staphylococcal nuclease to reduce the viscosity of the bioprocess feedstock. Biotechnology and Bioengineering. 2009;**104**:134-142

[34] Nesbeth D, Pardo MA, Ali S, Ward J, Keshavarz-Moore E. Growth and

productivity impacts of periplasmic nuclease expression in an *Escherichia coli* fab' fragment production strain. Biotechnology and Bioengineering. 2012;**109**:517-527

[35] Schofield DM, Sirka E, Keshavarz-Moore E, Ward JM, Nesbeth DN. Improving Fab' fragment retention in an autonucleolytic *Escherichia coli* strain by swapping periplasmic nuclease translocation signal from OmpA to DsbA. Biotechnology Letters. 2017;**39**:1865-1873

[36] Molin S, Givskov M, Riise E. Production in *Escherichia coli* of Extracellular Serratia spp. Hydrolase. Benzon Pharma, A/S, Hvidovre, Denmark. European Patent No. 0229866; 1992

[37] Albertsson PA. Partitions of Cell Particles and Macro-Molecules. New York: John Wiley & Sons, Inc.; 1960

[38] Philipson L, Albertsson PA, Frick G. The purification and concentration of viruses by aqueous polymer phase systems. Virology. 1960;**11**:553-571

[39] Lee YJ, Yi H, Kim WJ, Kang K, Yun DS, Strano MS, et al. Fabricating genetically engineered high-power lithium-ion batteries using multiple virus genes. Science. 2009;**324**:1051-1055

[40] Hales JE, Ward J, Aeppli G and Dafforn T. Fluorescent Composition. WO2013093499 PCT/GB2012/053236. 2013

Chapter 6

Surveillance and Elimination of Bacteriophage Contamination in an Industrial Fermentation Process

James A. Zahn and Mathew C. Halter

Abstract

Commercial fermentation processes are often vulnerable to bacteriophage due to the lack of genetic diversity and use of high cell density cultures. Bacteriophage infections in these fermentations can have adverse impacts on operability of the production facility and product quality and prevent recovery of valuable bioproducts in the downstream process. Prevention strategies have been developed and optimized through feedback from bacteriophage diagnostic tests, which inform improvements to process design for elimination of entry points, as well as modification of the biocatalyst to reduce or eliminate bacteriophage virulence. In this chapter, we provide case studies for successful elimination of bacteriophage virulence via host modifications, including bacteriophage binding-site modifications on the outer membrane of an *Escherichia coli* production host, used for commercial manufacture of 1,3-propanediol, as well as application of CRISPR-associated protein 9 (Cas9) for bacteriophage immunity. Finally, we report application of bacteriophage diagnostic methods to fully characterize and eliminate bacteriophage entry points in a commercial fermentation process.

Keywords: bacteriophage, white biotechnology, industrial fermentation, CRISPR, Cas9, phage

1. Introduction

Increasing awareness of our dependence on petroleum, coupled with the negative effects of this dependency on oil supply, price volatility, and gas emissions, is the driver behind the growing global market for biorenewable chemicals made through white biotechnology processes. In 2011, revenue from biorenewable chemicals exceeded $2.4 billion, and revenue continues to grow at a compound annual growth rate of 14.8%, with glycerin and lactic acid accounting for 79% of the market share [1]. Growth for biorenewable chemicals that are used as monomers the manufacture of bioplastics represents a fast-growing segment of this industry [1].

One example of a biorenewable chemical that has displaced petrochemical manufacturing routes is 1,3-propanediol or BioPDO™. This chemical had historically been manufactured using petroleum-derived ethylene oxide or acrolein [2, 3]. In 2006, DuPont Tate & Lyle Bio Products, a joint venture formed between DuPont and Tate & Lyle, commercialized an aerobic fermentation process for production of 1,3-propanediol (BioPDO™) from glucose derived from yellow dent field corn.

IntechOpen

A life cycle analysis (LCA) showed that the BioPDO™ manufacturing process consumed 42% less energy and emitted 56% less greenhouse gas emissions than petrochemical manufacturing routes. BioPDO™ is included as a raw material in polyester manufacture for textiles, carpet (Sorona®), and thermoplastic resins (Sorona®EP). Furthermore, BioPDO™ has direct uses in foods, cosmetics, and personal care applications through the Zemea® brand, and heat transfer, homopolymer, polyurethane, and other industrial applications through the Susterra® brand [3, 4].

In addition to a more favorable life cycle analyses, White Biotechnology processes like the BioPDO™ process demonstrate many advantages over petrochemical processes, such as the ability of the process to maintain high specificity [5], high yield, and ability to maintain lower concentrations of chemical intermediates that are generally recognized to form undesirable impurities in the final product [6].

Although many advantages exist for bioprocesses, disadvantages include (a) low to moderate productivity rates requiring intensification capital investments related to equipment capacity, (b) process susceptibility to bacterial and bacteriophage contamination, which has a negative impact on operability factor, (c) generally higher variable cost of manufacturing due to high electrical and related energy costs, (d) high water use, (e) odors, (f) costs associated with the production and filtration of large volumes of gases required for the fermentation process, and treatment of waste gases to remove odors, viable cells, and regulated or unregulated chemicals, (g) higher separation and refining costs due to the presence of large amounts of water, and (h) disposal of large volumes of nonhazardous waste.

In this chapter, we provide a detailed analysis of one of the most challenging issues for white biotechnology processes, which is the prevention of lytic bacteriophage events in commercial fermentors. Bacteriophages can cause rapid lytic infections of the highly clonal bacterial populations that are used in white biotechnology processes. Lytic infections of bacteria by bacteriophage reduce or abolish product productivity, and reduce the efficiency of cell recovery methods, which causes reduced product quality or a complete loss of the product. Batches affected by lytic bacteriophage infections cause a loss in production capacity or asset utilization, and financial losses to a business [7]. Lytic events in industrial fermentation may necessitate temporary shutdown of the facility for cleaning and elimination of bacteriophage, or even prolonged shutdown periods for cleaning and modification of aseptic barriers in the facility. In addition to surveillance and elimination best practices, this chapter outlines facility design considerations that are important in the prevention of bacteriophage in a biomanufacturing facility.

2. Prevalence of bacteriophage contamination in industrial fermentation processes

With the steady increase in the use of prokaryotic biocatalysts over the course of the last several decades for protein, small molecule, and chemical production, a focus has been placed on maintaining a bacteriophage-free environment in the manufacturing facility. The prevalence of bacteriophage in commercial fermentation processes varies considerably within the industry. For example, there have been no reported bacteriophage-related losses in certain industrial process, including: (a) syngas fermentations utilizing *Clostridium ljungdahlii*, *Clostridium autoethanogenum*, and *Clostridium coskatii*; (b) the commercial process for production of Spinosad and Spinetoram using *Saccharopolyspora spinosa*; and the commercial process for (c) production of xanthan gum by *Xanthomonas campestris*. In contrast, moderate to severe fermentative losses have been observed for certain bacterial-mediated processes, including the acetone-butanol-ethanol process that uses *Clostridium acetobutylicum*, *Clostridium beijerinckii*, or *Clostridium* sp. [8]; the *Escherichia coli*-mediated NutraSweet® process [9]; the *E. coli*-mediated BioPDO™ process [10]; the vinegar fermentation processes that uses *Acetobacter europaeus*, or *Acetobacter* sp. [11]; and food processes that utilize various bacteria [12, 13].

Although no single factor appears to link processes that are susceptible to bacteriophage-induced fermentative losses, some common themes exist:

a. Tanji et al. [14] proposed that "the most likely source of phage contamination in the *E. coli* culturing process is human, since *E. coli* is one of the main inhabitants of the gastrointestinal tracts of warm-blooded animals." Infection of microflora present in fermentation plant personnel may serve as a reservoir to maintain bacteriophage in the plant environment for extended periods of time [15]. Production hosts that are incapable of human colonization or that exhibit distinct phylogenetic differences to the natural human microflora may be advantaged over *E. coli* production hosts for these processes.

b. Certain processes using *E. coli* have a distinct disadvantage that aseptic design and operation principles are unknowingly sacrificed by continuous improvement programs, which focus on minimizing process cycle time, or related cost reductions. Processes that produce end products that are inhibitory to foreign bacterial growth are especially vulnerable since a false sense of security is provided by the fact that the *E. coli* production hosts exhibit improved competitiveness over most other microorganisms. Under these conditions, foreign growth, which may harbor bacteriophage, can go undetected or unreported. Process cycle time improvements typically target a reduction in maintenance tasks supporting sterile barriers, or a reduction in time allotted for sterilization and clean-in-place (CIP) procedures to clean piping, valves, and vessels. These efforts can negatively impact bacteriophage elimination programs.

c. *Xanthomonas campestris* has a number of established bacteriophages that have been registered as biocontrol agents to minimize its impact as a plant pathogen [16]. Therefore, the absence of bacteriophage-induced losses for the commercial xanthan gum process is aligned with the absence of an appropriate reservoir in the manufacturing facility that ecologically supports *X. campestris* growth.

d. Microorganisms with doubling times greater than 2–3 hours, which includes *Saccharopolyspora spinosa*, *Clostridium ljungdahlii*, and *Clostridium autoethanogenum*, often exhibit effective DNA restriction-modification systems that serve to destroy foreign DNA that enters the cell. These microorganisms are typically referred to as recalcitrant to recombinant DNA technology, as measured by low or unmeasurable transformation efficiency [17]. More recently, a number of widespread bacteriophage immunity systems have been described, some of which contain elements that act similarly to restriction-modification systems, to modify or destroy both foreign (i.e., non-native plasmid or cosmid-associated DNA) and bacteriophage-associated DNA [18, 19]. Interestingly, the *E. coli* production host used for the BioPDO™ processes was found to lack a functional bacteriophage exclusion operon (BREX [18]), as well as a functional CRISPR-Cas restriction-modification system [7, 20]. Highly optimized strains utilized in commercial manufacturing process often contain intended and unintended modifications to the host chromosome that contribute to improved rate,

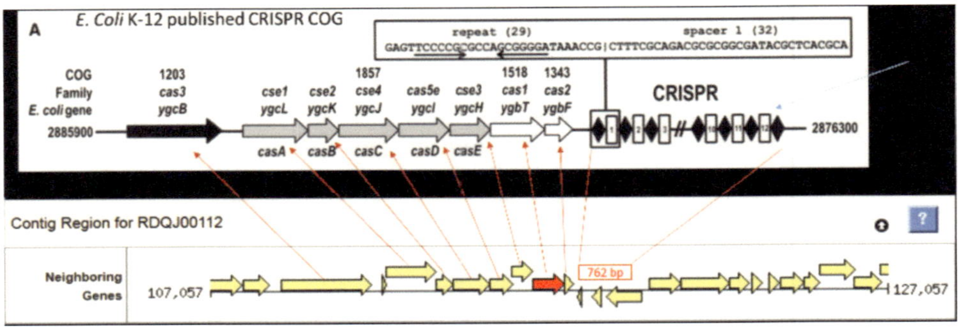

Figure 1.
Alignment of the E. coli K12 CRISPR operon (upper panel) with the native nonfunctional CRISPR-Cas operon in the BioPDO™ E. coli production host (lower panel).

titer, and yield. Although the BioPDO™ production host contains nearly a complete *E. coli* K12 CRISPR-Cas operon (**Figure 1**), upstream regulatory modifications, which were correlated with improved product production rate resulted in the loss of function for this operon. Vale et al. [19] found that "Cas protein expression is particularly costly, as Cas-deficient mutants achieved higher competitive abilities than the wild-type strain with functional Cas proteins." Smaller spacer libraries of approximately four spacers or less were not associated with fitness costs, suggesting that the genetically engineered spacer library approach of Halter and Zahn may serve to minimize energy and ATP burden on the cell, as indicated by the minimal impact of recombinant seven-spacer system on 1,3-propanediol biosynthesis [7].

2.1 Isolation and identification of bacteriophage DNA from lytic production samples in the BioPDO™ process

Lytic bacteriophage infection with the BioPDO™ process is characterized by sudden cellular lysis, which coincides with a sudden and rapid increase in dissolved oxygen (to 100% dissolved oxygen), a complete loss of oxygen uptake rate, and a complete loss of carbon dioxide evolution rate during the fed-batch fermentation. Optical density (OD) of the fermentation decreases rapidly from an OD_{550} nm at approximately 11 hours of 42 ± 2 absorbance units to less than 1.4 ± 0.5 in a period of 30–40 minutes. Viable cell counts for the process show that a nearly complete 10-log reduction occurs within this time period. As in-house molecular techniques for bacteriophage detection were not initially developed at the time of facility start-up, these early events were poorly characterized. However, procedures existed to sample and preserve fermentor samples in segregated freezers to support future investigative efforts. Fermentor samples that were collected consisted of a crude mixture of cleared (lysed) *E. coli* cells, nucleic acids, proteins, and cell membrane components. These samples were subsequently determined to contain bacteriophage particles, which were directly observed by transmission electron microscopy (**Figure 2**).

2.1.1 DNA isolation methodology

Samples were syringe-filtered through a 0.2-μm filter to separate the larger cell debris, and bacteriophage particles were precipitated and concentrated for transmission electron microscopy (**Figure 2**). To improve DNA sequencing efforts,

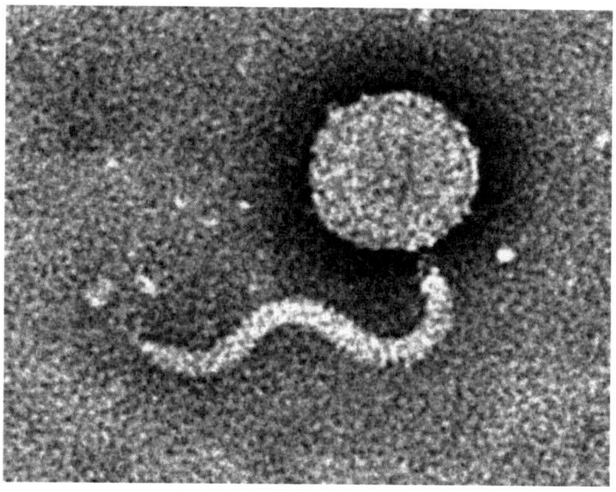

Figure 2.
Transmission electron micrograph of bacteriophage DTL.

bacteriophage nucleic acids were separated from lysed *E. coli* DNA. The bacteriophage DNA is protected by the proteinaceous capsid head, allowing for DNAse digest of exogenous *E. coli* DNA. Filtered samples were incubated with DNAse I according to the manufacturers recommended protocol overnight, ensuring complete digestion of all DNA present that could interfere with bacteriophage DNA sequencing. After complete digestion, the DNAse was deactivated by heat denaturation. The next step was the removal of the bacteriophage capsid protecting the DNA. This was performed by treatment with proteinase K according to the manufacturer's protocol for 3 hours. The proteinase K treatment removes the capsid protein shell, allowing the bacteriophage DNA to enter solution. Now that the bacterial DNA was fully degraded and the bacteriophage DNA removed from the particle capsid, the final remaining steps were simply precipitating the remaining protein from solution, separation from the aqueous phase by centrifugation, and precipitation of nucleic acids. Protein precipitation was performed by addition of 2 M potassium hydroxide. Immediate flocculent formation was evident, but the samples were stored on ice for a short period to encourage further protein precipitation. Chilled samples were centrifuged in a tabletop centrifuge at 13,000 × *g* for 10 minutes to separate the protein flocculent from the nucleic acid-containing supernatant. Upon separation, nucleic acids were precipitated through a 50% isopropyl alcohol wash. Samples were again stored on ice to allow precipitation to progress more efficiently, followed by centrifugation at 13,000 × *g* for 5 minutes. The visible, cloudy DNA pellet was resuspended in ddH_2O and quantified for sequencing.

2.1.2 Bacteriophage typing

The isolated DNA was sent for 454 pyrosequencing, which revealed an approximately 46,000 base pair circular genome. The genomic characterization of the plasmid was described previously [7]. Sequence analysis of genomes isolated from lytic batches ranging back to the manufacturing facility start-up in 2006 revealed an interesting fact that every lytic bacteriophage event was caused by the same bacteriophage. This bacteriophage shared sequence homology with a T1-like bacteriophage, referred to as RTP bacteriophage [21], but sequence homology of tail fibers and other open reading frames in the genome indicated it was a new representative of this group [7]. To reflect the source of the novel bacteriophage, it was named "*DTL*-phage," for the location of its discovery at the DuPont Tate & Lyle Bio Products fermentation manufacturing site. It is noteworthy that with the significant diversity of coliphages in nature, only one specific type of bacteriophage was detected over a period of 12 years. The prevalence of this particular bacteriophage at this facility was likely enhanced by the lack of significant changes in bacteriophage resistance in the *E. coli* host strain, and the massive amount of bacteriophage particles that was released from the site's five—600,000-L production fermentors. With a burst size of approximately 300 new phage particles with each cell lysis, infected fermentors were estimated to release approximately 2×10^{15} viral particles per affected fermentor, which were then distributed to vent and broth deactivation systems where these phage particles were presumed to be deactivated. Subsequent viral load studies showed that large amount (40–65%) of the bacteriophage present in liquid and respiratory off-gas analysis streams remained active. Although a majority of flow in these systems was contained, broth from in-process fermentor sampling efforts, and a majority of excess flow from off-gas analysis streams for respiratory gas analysis was not contained, and contributed to bacteriophage-containing aerosols, and dried particulate matter, which was disseminated in the production facility.

While we have shown that bacteriophage *DTL* is highly selective for *E. coli* (**Figure 3**), it is capable of lysing several different *E. coli* Group A representatives and related subspecies (**Figure 4**), suggesting that bacteriophage particles could infect related *E. coli* subspecies present in the local manufacturing environment, since these representatives are common human and animal inhabitants [14].

Since the bacteriophage plaque assays provide slow (34–48 hours) feedback for the presence of bacteriophage contamination in process samples, development of a rapid diagnostic test was considered a critical path for effectively mitigating fermentative losses in the facility. To this end, polymerase chain reaction (PCR) assays were developed to detect *DTL*-phage nucleic acids in fermentation broth (**Table 1**). Primer sets were designed based on the genome sequence ([22]; Genbank accession number MG050172), and endpoint reactions performed on both lysed and nonlysed fermentation samples were run on 1% agarose gel, providing a binary visual confirmation of the presence or absence of *DTL*-phage genomic DNA (**Figure 5**).

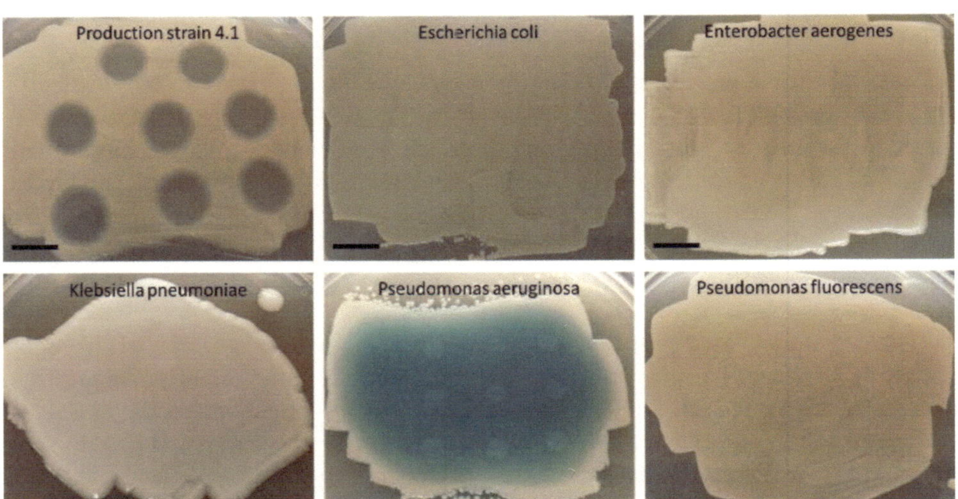

Figure 3.
Plaque assays using DTL-phage on phylogenetically diverse bacteria. From the top panel—left to right: Escherichia coli FM5, E. coli ATCC 8739, Enterobacter aerogenes ATCC 13048: from the lower panel—left to right, Klebsiella pneumoniae ATCC 15574, Pseudomonas aeruginosa ATCC 9027, and Pseudomonas fluorescens ATCC 13525. Background spotting, which is most notable with P. aeruginosa is due to nutrient carry over in the phage preparation.

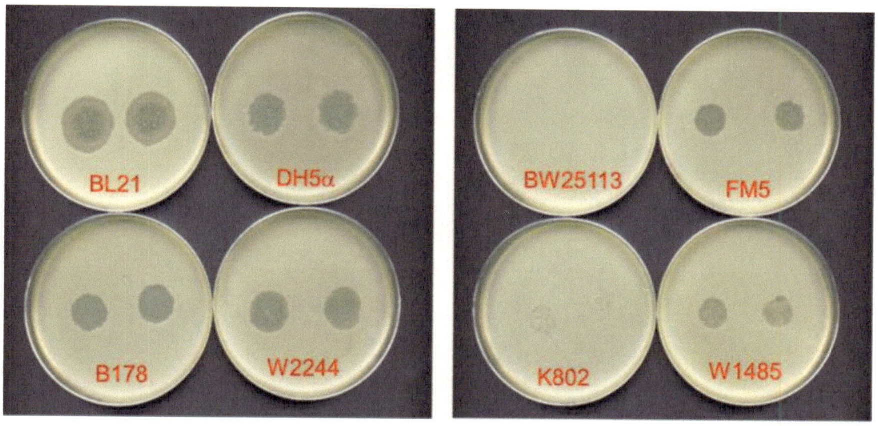

Figure 4.
Susceptibility of Escherichia coli group A representatives and related E. coli subspecies to DTL-phage. Susceptible = BL21, DH5α, B178, W2244, FM5, and W1485. Nonsusceptible = BW25113 and K802.

Primer	Sequence
DTL-phage 43 qRT F	GAAGAGGTGTTTAATTTCGCCG
DTL-phage 43 qRT R	GCCAAACCCGTTAATGTGAAC
DTL-phage 40 qRT F	AGAGGTAGTGGTAGGTTCCG
DTL-phage 40 qRT R	TCAAGAATCGCAGAGTAACCG
DTL-phage 19 qRT F	GCACGCTGGTTAATGGAATG
DTL-phage 19 qRT R	TTCTTGATGGAGATTGTCGGG
DTL-phage 6 qRT F	GGCGTAAAACAGTAATTCAGGTC
DTL-phage 6 qRT R	TTCACACCATCACTACCATCAG
DTL-phage 66 qRT F	GCAGTAAGCCAGAGATTAGCG
DTL-phage 66 qRT R	CTATCCAGTGACCCAACCTTG

Table 1.
DTL-phage primer sets for PCR diagnostic testing.

After optimization of the assay, early PCR screening of every fermenter was implemented in the manufacturing facility for transfer of seed fermentors. It was soon clear that endpoint PCR, followed by gel-electrophoresis was still too slow to effectively meet the 2-hour transfer window in the manufacturing process, and secondly, the technical tasks and decisions were too complex to effectively implement for the nontechnical manufacturing facility workforce. Real-time quantitative PCR is an assay that provides a more rapid result (<1 hour), and also produces a threshold cycle (C_t) count that can be used as a straightforward quality control parameter to set a simple quantitative threshold for amplicon copy numbers. The use of real-time quantitative PCR to track fermenter contamination provides a more rapid screening procedure and reduces the time from fermentor sampling to a formation of a transfer decision by slightly over 1 hour. Primer sets were optimized to amplify targets within the bacteriophage *DTL* genome with above 95% efficiency, and premade reaction cocktails were also developed to minimize low-volume pipetting steps needed to be performed by nontechnical staff. This procedure had the following steps: (a) A seed fermenter sample would be collected by the fermentation technician and transferred into

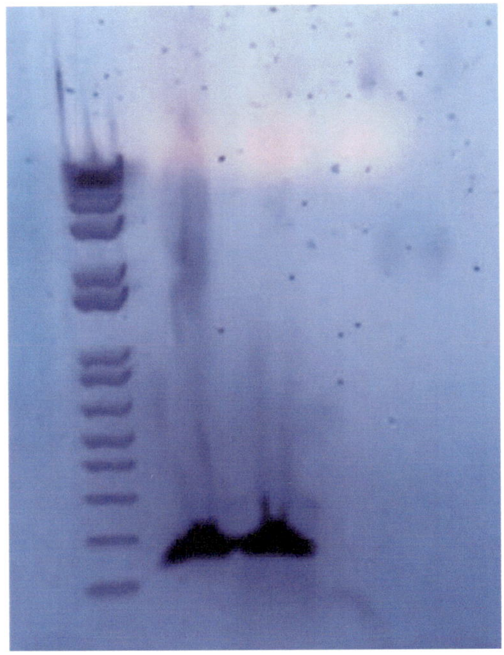

Figure 5.
DTL-phage amplicons generated from PCR (DTL-phage 40 set, **Table 1***) and separated by 1% agarose gel electrophoresis. Lane 2 is a positive control of purified DTL-phage DNA, lane 3 is a seed fermenter sample containing 2 × 10*^{4} *particles of DTL-phage per mL, and lane 4 is a sterile water negative control. Lane 1: Invitrogen E-Gel 1 Kb Plus Express DNA Ladder (part #10488091) 100–5000 bp.*

a class 2 biological safety cabinet. (b) The sample would be filtered through a 0.2-μm sterile syringe filter to remove intact cells, and then diluted 1:10 in sterile water. This dilution was performed to dilute inhibitory substances present in the fermentation media, allowing for elimination of the DNA extraction step. (c) Load the processed sample with positive and negative controls into the quantitative PCR instrument, and initiate the analysis. This qPCR assay was the cornerstone in a strategy to screen seed fermentors prior to transfer to the production fermentor as a safeguard to limit bacteriophage-infected seeds from contaminating production fermentors. This new procedure included control charting intensities for C_t trends, which aided in the tracking of bacteriophage contamination levels (**Figure 6**), which increased the operability factor for the

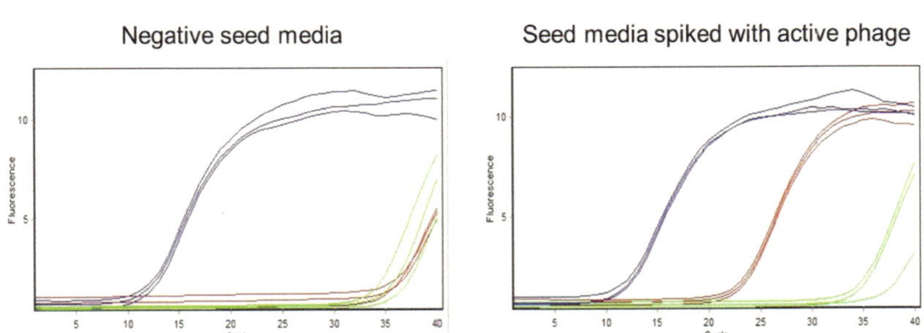

Figure 6.
Quantitative PCR (qPCR) amplification curves of seed media screened for DTL-phage. Blue colored curves represent a positive control for presence of DTL-phage (equivalent to 180 plaque forming units per milliliter), green curves represent a negative control, and red curves represent the seed media being screened. The left panel shows a negative result for the seed fermentation sample with no detectable DTL-phage, and the right panel shows a positive response for a seed fermentation sample that contained an internal addition of 80 plaque forming units per milliliter.

fermentation process by ensuring that only bacteriophage-free seed material was transferred into production fermentors.

2.2 Efforts to produce a bacteriophage-resistant production strain

There is a clear incentive for businesses that utilize fermentation-based manufacturing technologies to invest in the development of production strains that are resistant to potential bacteriophage threats. There are at least four target areas that have been described in literature for enhancing bacteriophage resistance in prokaryotic organisms that are sensitive to lytic bacteriophage infections [23]: (a) prevention of phage adsorption, (b) preventing phage entry, (c) cutting bacteriophage nucleic acids, and (d) abortive infection systems. The following analysis will summarize efforts in two of the four areas and prevention of phage adsorption and cutting phage nucleic acids with a heterologous CRISPR-Cas system.

Classical strain improvement programs for the generation of bacteriophage resistance often start with acquiring genotypic diversity in a host population through spontaneous or induced mutation, followed by challenge and selection of survivors to bacteriophage infection. Several studies utilizing this strategy have reported resistance through single nucleotide polymorphisms (SNPs) or insertional mutagenesis, which often alters the structure of bacteriophage binding sites on the cell surface. Additionally, significant deletions of chromosomal DNA have been described that involve structural changes of one or more cell surface biomolecules [14, 24–26].

The initial step in the bacteriophage virulence cycle is the adsorption of the bacteriophage particle to the outer surface of the bacterial cell [23]. Once bound, the bacteriophage infects the cell by transferring genetic material into the cell, where it then utilizes host cellular translation systems to generate additional bacteriophage particles. This binding of the bacteriophage to the outer membrane is typically mediated by a highly specific receptor on the outside of the cell, is typically a membrane protein, a specific class of lipid, or a carbohydrate moiety, and is extremely specific to a bacteriophage and its host. Common genetic techniques can, therefore, be used to target these binding sites to reduce binding affinity or eliminate the site [24, 26].

Over several years, classical strain improvement programs for the BioPDO™ process have generated and screened strains that were selected through challenge for resistance to *DTL*-phage [7]. These screens included strains that demonstrated a 3–5-log improvement in the reduced sensitivity as measured by the standard bacteriophage plaque assay (**Figure 7**). In all cases, bacteriophage resistance was associated with a 63–96% reduction in 1,3-propanediol production titer. Further analysis showed that these strains had a lower solvent tolerance, which indicated that the bacteriophage resistance mechanism(s) were presumably linked to changes in cellular ultrastructure. Genomic sequencing efforts confirmed that SNPs occurred in a series of genes involved in synthesis and glycosylation of lipopolysaccharide, including glycosylation with heptose residues [27]. Heptose residues appear to be essential for high-affinity binding of *DTL*-phage to the *E. coli* cell [28].

The second strategy, cutting bacteriophage nucleic acids is a rapidly evolving field that has been heavily influenced by recent discoveries in bacterial-acquired immunity [12]. Clustered regularly interspaced short palindromic repeats (CRISPR) are a molecular system by which prokaryotes obtain acquired resistance to bacteriophages. The CRISPR operon is widely distributed in prokaryotes and represents the most abundant form of innate immunity in these organisms [29]. Upon injection of the genetic material, the cell recognizes foreign DNA, and the first gene involved

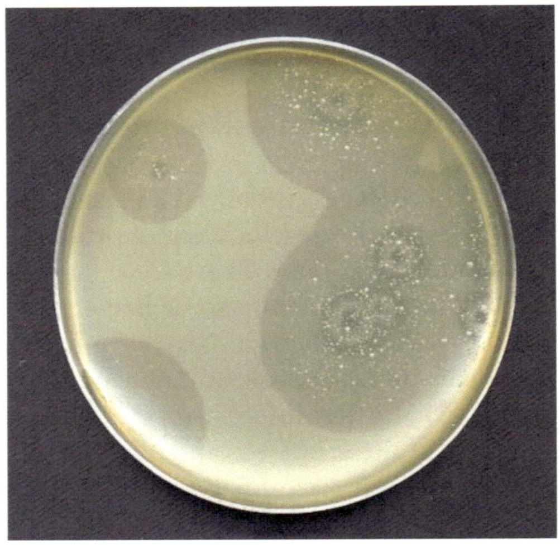

Figure 7.
Bacteriophage plaque assay plate showing the presence of E. coli colonies growing in the zone of clearing that are resistant to DTL-phage challenge.

in the CRISPR pathway creates an approximately 20–25 base pair cut (this varies between species and CRISPR subtypes). This snippet of DNA is then integrated into a library of recognized bacteriophage genetic elements, flanked on each side by a palindromic repeat. This genetic element can then be transcribed, producing a single-stranded transcript that is homologous to the DNA of the intruding bacteriophage. The palindromic repeats flanking the element serve to bind the short fragment of transcript to a nuclease, which is then guided by the homologous element to potential binding sites on the intruding segment of DNA, where a single cut is made rendering it nonfunctional.

Since it was established earlier that the BioPDO™ *E. coli* production host did not carry a functional CRISPR operon in its genome, and the specific binding site of the *DTL*-phage was originally unknown, work was initiated with CRISPR to generate a spacer library specific to *DTL*-phage infection. The well-characterized CRISPR cassette of *Streptococcus thermophilus* was amplified, and its 12 bacteriophage spacers (remnants of previously acquired resistance in this species) were replaced with seven spacers that were homologous to different regions of the *DTL*-phage genome. These seven spacers were specifically chosen to deliver CRISPR-nuclease-delivered cuts in the middle of important open reading frames within the bacteriophage genome, thereby ensuring overwhelming and targeted knockout of its genetic functionality.

Upon incorporation of this tailored cassette to the BioPDO™ *E. coli* production strain, plaque assays and phage challenges indicated the organism had obtained full resistance to this bacteriophage, as indicated by the lack of plaques formed in experiments using phage titers as high as 10^6 phage particles per milliliter. The utilization of multiple homologous spacers was an important aspect of this project to reduce the potential for resistance. Because CRISPR resistance depends on homologous binding of spacers bound to the CRISPR nuclease to the intruding phage DNA, a single SNP acquired within the homologous stretch of DNA in the phage genome could disrupt the DNA-nuclease interaction, rendering that spacer no longer active. The use of seven targeted spacers minimizes the potential for resistance, and no significant energy or ATP burden related to the lack of fitness hypothesis, proposed by Vale et al., was observed [19]. Although Vale et al., proposed a threshold number of four spacers as the limit regarding negative impacts on cell fitness, the slight increase to a seven-spacer library appeared to have no negative consequences on 1,3-propanediol synthesis rate [7].

2.3 Reservoirs and process entry points for bacteriophage in a manufacturing setting

Host resistance to bacteriophage infection is only a partial solution to the prevention of bacteriophage in fermentation processes. This approach must be balanced with effective facility sanitation procedures, active bacteriophage surveillance programs that are coupled with reservoir elimination efforts, and finally, the validation, maintenance, and monitoring of aseptic barriers utilized in the process for bacteriophage exclusion.

As mentioned previously, bacteriophage-infected production fermentors for the BioPDO™ process were estimated to release approximately 2×10^{15} viral particles per affected fermentor based on a 300-burst size model. Bacteriophage-contaminated air and liquid in fermentors was transferred to vent and broth deactivation systems where the viral particles were presumed to be deactivated. Earlier pilot plant studies of microbial and biological materials deactivated by a high-temperature short-time sterilizer (HTST) did not consider bacteriophage inactivation because (a) bacteriophage was considered an addressable nuisance in the operation of the facility, (b) bacteriophage libraries specific to the BioPDO™ production microorganism had not been assembled, and (c) risks to a commercial process remained largely speculative. Furthermore, the closest-related bacteriophage, *RTP*-phage had not been isolated and described until after the plant start-up in 2006. Subsequent heat-inactivation studies on phage-contaminated batches after facility start-up were successful in showing that bacteriophage particles were not completely deactivated in the HTST broth deactivation system, mainly due to resistance to heat inactivation [30, 31].

Other weaknesses in containment systems were gas and liquid sampling systems, which were commonly used in the fermentation industry for respiratory gas analysis and control through feedback loops, and in-process analytical for measuring broth parameters associated with growth and product production, respectively. Once free of containment systems, the bacteriophage particles were unintentionally disseminated by human activity and other environmental mechanisms. This disseminated material served to challenge sterile barriers that were in place for air filtration, filters used for sterile filtration of liquids used in the process, and cross contamination of microbiology process laboratories, which impacted the seed train, and laboratory diagnostic tests.

The ecological aspects of DTL bacteriophage are poorly characterized due to the lack of historical environmental samples from the fermentation facility during the period of the initial infection. We are in agreement with the proposal of Tanji et al. [14] that the most likely source of phage contamination for an *E. coli* culturing process is human, and the initial source of DTL-phage in the BioPDO™ facility, was most likely also human. The fact that DTL phage is now prevalent in the plant environment is thought to be mainly a result of a series of lytic events in production fermentors that established localized concentrations of the bacteriophage in reservoirs that are inhabited by natural populations of *E. coli*, or closely related microorganisms that are susceptible to this bacteriophage.

One of the most significant failure points in the commercial fermentation facility process for bacteriophage entry was through sterile filtration barriers. The small size of DTL-phage (capsid diameter = 0.075 μm) supports passage of the virus through liquid-service filtration systems (0.2–0.45 μm cutoff size), and conditional passage through gas-phase filtration systems (0.2 μm cutoff size) when entrained liquid is present. Aerial transfer of bacteriophage particles is well-established as a route for contamination of industrial fermentation processes [29, 30]; therefore, air filtration systems play a large role in protecting the fermentation process from airborne bacteriophage particles.

Our empirical studies on the Donaldson sterile air filter, model P-SRF N 30/30 (0.2 μm absolute, expanded PTFE filter membrane) indicated that the filter was 100% effective in rejection of DTL-phage particles if the filter remained above the system pressure dew point temperature of 102.5°F (**Figure 8**). Since the design temperature of air present in the header was 122, the 102.5°F pressure dew point was considered to have sufficient safety factor to prevent formation of condensate. Wetting of the filter was correlated strongly with lytic infections by DTL-phage [7]. Wetting of the sterile filter due to entrained water in the process air header was infrequently observed when: (a) low ambient temperatures cooled areas of the process air header that were not adequately protected with heat tape and/or insulation (**Figure 9**), (b) cooling tower water or chilled water from the intercooler and/or aftercooler heat exchangers on the air compressor leaked into the air stream and traveled through the air header to prefilters and sterile filters, (c) low-point zones in the process header not adequately drained with autopurging condensate traps, and (d) high-temperature excursions of the chilled water system, which was used to condense water in the compressed air product stream. A critical factor for surveillance and prevention of bacteriophage contamination was the integration of dew point sensors at upstream locations from the sterile filter, and more frequent sterilizations of the filters in the sterile filter housing. Since sterile filters have a limitation of approximately 170 steam cycles, more frequent sterilization served to increase the number of filter replacements and overall process costs.

Bubble column bioreactors used in aerobic fermentation processes require large volumes of air to meet the oxygen uptake requirements for *E. coli* cells in the fermentation, as well as a means for mixing to drive heat transfer, gas solubility, CO_2 ventilation or removal, and mixing of substrates, and other feeds. Due to the expense of water vapor removal in very large volumes of air, many fermentation systems utilize ANSI/ISA-7.0.01-1996 air quality class 6 or lower specifications, which establish a dew point of 50°F at atmosphere pressure. This type of air compressor system, which contains chilled fluid after-coolers, passes the compressed air through a cooled heat exchanger, causing water vapor to condense out of the air stream, and typically produces air with a dew point not lower than 41°F (5°C). The pressure dew point refers to the dew point temperature of a gas under pressure, which in this case is increased to a value of 102.5°F at a pressure of approximately 84.7 psia (**Figure 9**). The main issues with this system are (a) the potential for

Figure 8.
Sterile process of destructive disassembly a Donaldson sterile air filter, model P-SRF N 30/30 (0.2 μm absolute, expanded PTFE filter membrane) for bacteriophage diagnostic testing. Note the dark water staining on protective prefilter barrier layer on the filter.

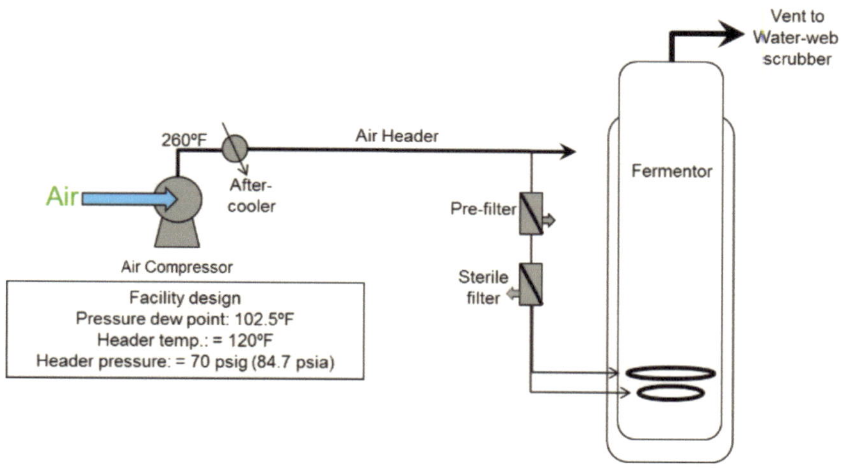

Figure 9.
Diagram of the process air system that provided a sterile-filtered air to a production fermentor.

heat exchanger leakage into the air stream and (b) temperature excursions of the cooling fluid (chilled water), due to its use in other areas of the manufacturing facility. Temperature excursions for chilled water were found to reduce the ability of the after-cooler to remove water vapor in air supplied to the fermentors, which increased the potential for condensate to form in the process air header. This condensate was found to cause passage of bacteriophage particles through the sterile, 0.2-μm air filter.

Deviations from aseptic design principles are nearly always a root cause for introduction of foreign bacterial growth and bacteriophage into axenic fermentation systems. A deviation in these principles occurred as part of the risk mitigation effort to prevent condensate formation in the process air header. As part of this project, sterile air bypass lines were installed on the air header line to eliminate zero air flow conditions that occurred during the sterilization cycle for fermentors. During fermentor sterilization, sterile air flow was stopped, and a 45 psig steam feed line between the sterile filter housing and fermentor was opened. During the 90-minute sterilization cycle, the temperature of the air header was found to drop well below the pressure dew point, especially during cold weather months, which introduced condensate upstream of the sterile air filters. Introduction of a process steam line upstream of the sterile filter housing, as is done more commonly in the industry, was not considered a best practice because the sterile air filter would remain wetted for a short period after the steam sterilization process was completed, and nonsterile air flow was reinitiated as a feed stream to the filter housing. The wetted filter was believed to promote passage of bacteriophage particles into the fermentor. One solution, practiced at a separate facility, was to have a secondary low-flow sterile air header upstream of the main sterile filter housing that provided sterile dry air to the housing after the steam sterilization to dry the wet sterile filters. As these former solutions required significant down time to perform repiping, two other solutions were prioritized: (1) ensure that heat tracing and insulation of the air header was sufficient to maintain temperature above the pressure dew point and (2) prevent the low or no-air-flow condition by adding a small diameter purge line directly after the sterile air filter housing to provide constant flow in the air header. Due to the potential for ammonia gas in the bypass air stream, piping effort focused on connection of this line to the fermentor exhaust header, which contained an ammonia scrubber system. The new piping system did not contain proper aseptic barriers and steam-purged interlocks, which then permitted backflow of contaminated waste gas into the fermentor air header (poststerile filter) under infrequent vent header over-pressure events.

One of the more surprising discoveries in this work was the level of bacteriophage contamination (up to 7×10^3 plaque forming units per square centimeter) in the process air header, as well as the surface of sterile air filters between sterilization cycles. This result indicated that the extremely short, but elevated air temperature of 260°F between the air compressor and after-cooler was insufficient in deactivating *DTL*-phage (**Figure 9**). Based on this finding, air header sanitation procedures were developed to infrequently steam sterilize the air header for the location between the preair filter housing and the air compressors. While this practice reduced phage burden in the air header, it could not be operated in a continuous mode, and therefore, the sterile air filters served as the only barrier to bacteriophage entry. Additional design improvements related to a secondary barrier for airborne phage entry include: (a) a heat-sterilization system on the product side of the air compressor to elevate the temperature to 300°F, and increase temperature hold time by 10-fold to 1 second, and (b) filtration/precipitation systems for pretreatment of supply air to the air compressors.

Generation of bacteriophage aerosols in the BioPDO™ manufacturing facility is largely driven by failure points in containment systems, including the industry-standard water-web hypochlorous acid scrubber system for fermentor vent stream treatment, as well as the use of open liquid transfer systems, such as trench drains that are used for streams originating from steam interlocks on sample systems, drains from sample sinks, drains from vapor–liquid separators, and unintended leaks from heat exchangers and piping. Efforts to ensure that these streams are transferred by closed piping systems to chemical or steam deactivation systems is critical for achieving low airborne concentrations of bacteriophage in the manufacturing environment. Frequent cleaning of these areas with disinfectants has been shown to be critical for limiting dispersion in the manufacturing facility environment [32].

3. Conclusions

As the most abundant biological entity on the planet, bacteriophages play an important ecological role in ecosystems and have also been exploited for the development of many modern technologies, including gene transfer and treatment of bacterial infections [7, 14, 21, 23]. The presence of bacteriophage in an industrial fermentation facility can be a serious problem, resulting in reduced product quality, loss in production facility operability, and financial losses to a business. In this chapter, we have described basic and applied research efforts around our goal of reducing the impact of bacteriophage-related losses in a commercial process for the manufacture of 1,3-propanediol. The key to success in these efforts was the development of rapid diagnostic methods that were subsequently leveraged by a diverse team to quickly diagnose and eliminate sources of bacteriophage in a fast-paced manufacturing environment.

Many of the solutions related to operating a manufacturing facility bacteriophage-free require participation of a cross-disciplinary team that encompass many areas of expertise, including virology, microbiology, microbial ecology, chemical, and mechanical engineering. As with any new biomanufacturing process, designers utilize a basic set of assumptions that serve in the design and construction of the facility. Some of these assumptions are not fully tested in a pilot plant, and the issues that arise are often seen for the first time in the commercial fermentation facility. In this chapter, we have characterized issues related to the use of 0.2 μm (absolute) sterile filters as the only mechanism for exclusion of bacteriophage from the fermentation process. This issue is especially problematic since these filters were not specifically designed for removal of irregular bacteriophage particles with diameters (0.075 μm), which is significantly smaller than the stated filter cutoff size.

Our studies indicate that CRISPR/Cas9-mediated resistance to bacteriophage DTL in the *E. coli* PDO production strain is a highly effective strategy for eliminating bacteriophage virulence. The CRISPR spacer customization strategy further ensures that spacers are not generated against foreign DNA that that is inserted into the production host as part of future host improvement (transformation) projects. Furthermore, the metabolic burden hypothesis proposed by Vale et al. [19] is avoided since we have limited the spacer library to a total of seven spacers. The main disadvantage of this approach is that this recombinant CRISPR system has a narrow range of bacteriophage specificity, and that there is a significant selection pressure for sequence modifications that are not recognized by this existing spacer library. In this regard, there is a difficult balance between spacer library size as an assurance of bacteriophage resistance, and acceptance of lower 1,3-propanediol productivity due to the metabolic burden due to the increase in spacer library size.

Conflict of interest

The authors declare that there are no conflict of interests with this work.

Author details

James A. Zahn* and Mathew C. Halter
DuPont Tate & Lyle Bio Products, Loudon, TN, USA

*Address all correspondence to: james.a.zahn@dupont.com

References

[1] Sanford K, Chotani G, Danielson N, Zahn JA. Scaling up of renewable chemicals. Current Opinion in Biotechnology. 2016;**38**:112-122

[2] Kraus GA. Synthetic methods for the preparation of 1,3-propanediol. CLEAN—Soil, Air, Water. 2008;**36**(8):648-651

[3] Urban RA, Bakshi BR. 1,3-Propanediol from fossils versus biomass: A life cycle evaluation of emissions and ecological resources. Industrial and Engineering Chemistry Research. 2009;**48**:8068-8082

[4] DuPont Tate & Lyle Bio Products. Life Cycle Analysis Overview–Susterra® Propanediol. 2016. Available from: http://www.duponttateandlyle.com/sites/default/files/Susterra%20 LCA.pdf

[5] Pauling L. Molecular basis of biological specificity. Nature. 1974;**248**:769-771

[6] Weider PR, Powell JB, Slaugh LH, Forschner TC, Semple TC. Process for Preparing 1,3-Propanediol. United States Patent Number 5,545,767. United States Patent and Trademark Office; 1996

[7] Halter MC, Zahn JA. Characterization of a novel lytic bacteriophage from an industrial *Escherichia coli* fermentation process and elimination of virulence using a heterologous CRISPR–Cas9 system. Journal of Industrial Microbiology & Biotechnology. 2018;**45**:153-163. DOI: 10.1007/s10295-018-2015-7

[8] Jones DT, Shirley M, Wu X, Keis S. Bacteriophage infections in the industrial acetone butanol (AB) fermentation process. Journal of Molecular Microbiology and Biotechnology. 2000;**2**(1):21-26

[9] Guiness R. 2006. Personal communication

[10] Nakamura CE, Whited GM. Metabolic engineering for the microbial production of 1,3-propanediol. Current Opinion in Biotechnology. 2003;**14**:454-459

[11] Sellmer S, Sievers M, Teuber M. Morphology, virulence and epidemiology of bacteriophage particles isolated from industrial vinegar fermentations. Systematic and Applied Microbiology. 1992;**15**:610-616

[12] Sturino JM, Klaenhammer TR. Engineered bacteriophage-defense systems in bioprocessing. Nature Reviews Microbiology. 2006;**4**:395-404

[13] Whitehead HR, Hunter GJE. Bacteriophage infection in cheese manufacture. The Journal of Dairy Research. 1945;**14**:64-80

[14] Tanji Y, Hattori K, Suzuki K, Miyanaga K. Spontaneous deletion of a 209-kilobase-pair fragment from the *Escherichia coli* genome occurs with acquisition of resistance to an assortment of infectious phages. Applied and Environmental Microbiology. 2008;**74**(14):4256-4263. DOI: 10.1128/AEM.00243-08 Epub 2008 May 23

[15] Bennett AM. Health hazards in biotechnology. In: Hambleton P, Melling J, Salusbury TT, editors. Biosafety in Industrial Biotechnology. Dordrecht: Springer; 1994. DOI: 10.1007/978-94-011-1352-6_7

[16] Jones JB, Jackson LE, Balogh B, Obradovic A, Iriarte FB, Momol MT. Bacteriophages for plant disease control. Annual Review of Phytopathology. 2007;**45**(1): 245-262

[17] Matsushima P, Broughton MC, Turner JR, Baltz RH. Conjugal transfer of cosmid DNA from *Escherichia coli* to *Saccharopolyspora spinosa*: Effects of chromosomal insertions on macrolide A83543 production. Gene. 1994;**146**(1):39-45

[18] Goldfarb T, Sberro H, Weinstock E, Cohen O, Doron S, Charpak-Amikam Y, et al. BREX is a novel phage resistance system widespread in microbial genomes. The EMBO Journal. 2015;**34**(2):169-183. Published online 2014 Dec 1. DOI: 10.15252/embj.201489455

[19] Vale PF, Lafforgue G, Gatchitch F, Gardan R, Moineau S, Gandon S. Costs of CRISPR-Cas-mediated resistance in *Streptococcus thermophiles*. Proceedings of the Biological Sciences. 2015;**282**(1812):20151270. DOI: 10.1098/rspb.2015.1270

[20] Tamulaitis G, Venclovas C, Siksnys V. Type III CRISPR-Cas immunity: Major differences brushed aside. Trends in Microbiology. 2017;**25**(1):49-61. DOI: 10.1016/j.tim.2016.09.012

[21] Wietzorrek A, Schwarz H, Herrmann C, Braun V. The Genome of the novel phage Rtp, with a Rosette-Like tail tip, is homologous to the genome of phage T1. Journal of Bacteriology. 2006;**188**(4):1419-1436. DOI: 10.1128/JB.188.4.1419-1436.2006

[22] Genbank accession number MG050172. *Escherichia* phage DTL, complete genome. Available from: https://www.ncbi.nlm.nih.gov/nuccore/MG050172

[23] Labrie SJ, Samson JE, Moineau S. Bacteriophage resistance mechanisms. Nature Reviews Microbiology. 2010;**8**:317-327

[24] Le S, Yao X, Lu S, Tan Y, Rao X, Li M, et al. Chromosomal DNA deletion confers phage resistance to *Pseudomonas aeruginosa*. Scientific Reports. 2014;**4**:4738. Published online 2014 Apr 28. DOI: 10.1038/srep04738

[25] Duerkopa BA, Huoc W, Bhardwajc P, Palmerc KL, Hooper LV. Molecular basis for lytic bacteriophage resistance in enterococci. MBio. 2016;**7**(4):e01304-e01316. DOI: 10.1128/mBio.01304-16

[26] Lucchini S, Sidoti J, Brüssow H. Broad-range bacteriophage resistance in *Streptococcus thermophilus* by insertional mutagenesis. Virology. 2000;**275**(2):267-277

[27] Benz I, Schmidt MA. Glycosylation with heptose residues mediated by the *aah* gene product is essential for adherence of the AIDA-I adhesion. Molecular Microbiology. 2001;**40**(6):1403-1413

[28] Gronow S, Brabetz W, Brade H. Comparative functional characterization in vitro of heptosyltransferase I (WaaC) and II (WaaF) from *Escherichia coli*. European Journal of Biochemistry. 2000;**267**:6602-6611

[29] Koonin EV, Makarova KS, Zhang F. Diversity, classification and evolution of CRISPR-Cas systems. Current Opinion in Microbiology. 2017;**37**: 67-78

[30] Sing EL, Elliker PR, Sandine WE. A method for evaluating the destruction of air-borne bacteriophages. Journal of Milk and Food Technology. 1964;**27**(5):125-128

[31] Jonczyk E, Klak M, Miedzybrodzki R, Gorski A. The influence of external factors on bacteriophages—review. Folia Microbiologica. 2011;**56**:191-200

[32] Martinez JE. The rotation of disinfectants principle: True or false. Pharmaceutical Technology. 2009;**33**(2):58-71. Available at: http://www.pharmtech.com/rotation-disinfectants-principle-true-or-false

Chapter 7

Targeting Peptides Derived from Phage Display for Clinical Imaging

Supang Khondee and Wibool Piyawattanametha

Abstract

Phage display is a high-throughput technology used to identify peptides or proteins with high and specific binding affinities to a target, which is usually a protein biomarker or therapeutic receptor. In general, this technique allows peptides with a particular sequence to be presented on a phage particle. Peptides derived from phage display play an important role in drug discovery, drug delivery, cancer imaging, and treatment. Phage peptides themselves can act as sole therapeutics, for example, drugs, gene therapeutic, and immunotherapeutic agents that are comprehensively described elsewhere. In this chapter, we discuss phage selection and screening procedures in detail including some modifications to reduce nonspecific binding. In addition, the rationale for discovery and utilization of phage peptides as molecular imaging probes is focused upon. Molecular imaging is a new paradigm that uses advanced imaging instruments integrated with specific molecular imaging probes. Applications include monitoring of metabolic and molecular functions, therapeutic response, and drug efficacy, as well as early cancer detection, personalized medicine, and image-guided therapy.

Keywords: peptides, membrane receptors, imaging, phage display, endoscopy

1. Introduction

One of the most important practices in modern era clinical imaging is imaging at the molecular level which can help characterize and measure in vivo biological processes at the cellular level [1]. Thus, the technique provides unambiguous and high-resolution real-time information for disease diagnoses and therapies. In addition, molecular imaging is usually noninvasive as a biologically active, receptor-specific, targeting vector conjugated to a radioligand, nanoparticle, and/or fluorescent/magnetic resonance imaging (MRI) probe is administered first, and then the probe signals can be quantified by positron emission tomography (PET), single-photon emission computed tomography (SPECT), magnetic resonance imaging (MRI), fluorescence imaging, or ultrasound. The specificity of probes may be contributed from targeting peptides, small proteins, and antibodies linked to the probes [2].

One of the most important aspects in successful molecular imaging is the development of imaging probes. Initial efforts focused on probes that are radiolabeled small molecules or macromolecules, e.g., monoclonal antibodies and their fragments [3]. Most such probes are unsuccessful because small molecules provide low specificity, whereas the antibodies have low target permeability. Taken into account altogether, these probes have low contrast between target tissues and background, leading to poor imaging qualities. Compared to small molecules and antibodies,

 IntechOpen

peptide imaging probes are more promising. The peptide length varies from several to approximately 50 amino acids [4]; thus they are usually more specific than small molecules and also more permeable than antibodies. The peptides have high capillary permeability, which allows efficient penetration into tissues. In addition, they also have high uptake rates in the target and rapid clearance from blood [1]. These distinctive advantages facilitate peptides as popular imaging probes. Such a probe is usually composed of a targeting peptide, a linker, and an imaging moiety. The linkers commonly are organic spacers, macrocyclic or branched chelators, and polymers, which link peptides with appropriate moieties. Different moieties render the probes observable by various devices, e.g., near-infrared (NIR) fluorescent dyes or quantum dots for optical imaging, radionuclides for PET or SPECT, and para-magnetic agents for MRI.

Phage display technology is a powerful approach to screen for peptides with high affinities and specificities to biomarkers. This technology was established by Smith et al. in 1985 to display polypeptides on the surface of filamentous M13-derived bacteriophage (phage) [5]. This technique modifies the phage genome to fuse the deoxyribonucleic acid (DNA) encoding a peptide to a gene encoding a protein comprising the phage coat; thus the peptide appears on the surface of the phage. In this way, each phage contains a single-peptide variant and its encoding DNA sequence, thus retaining a genotype-phenotype linkage. A library or pool of phages normally contains 10^9–10^{11} peptide variants for screening. The selection procedure consists of three main steps: (1) panning the pool of phages on the immobilized biomarker, (2) removing unbound phages by washing, and (3) elution of bound phages. After several rounds of such selections, the peptide sequences with high affinities to the biomarker are determined by sequencing the encoding DNAs in the phages. Specificity of the peptides can be also improved by adding extra negative selection steps [6].

Since its inception nearly 45 years ago, phage display has been widely used in thousands of research papers to isolate peptides that bind various targets [7]. The phage peptides are labeled with imaging agents such as radioactives, fluorescences, and nanoparticles. These probes have been successfully used to image tumors and cancers [8–10]. Moreover, phage display also influenced many other scientific fields such as drug discovery, vaccine development, and targeted drug delivery and gene therapy. With advances in molecular biology, the number of disease-associated biomarkers at the molecular level is ever-increasing. These new discoveries are motivating the applications of phage display to diagnostic imaging and targeted drug delivery.

2. Phage biology and phage selection screening methods

2.1 Phage biology

Bacteriophages (or phages) are viruses that infect bacteria. Phage virions vary widely in size, shape, and complexity, and phage genomes range in size from 3.4 kb to almost 500 kb [11]. Most phage genomes (>95%) discovered to date are linear, double-stranded DNAs (dsDNAs), but there are also genomes that are single-stranded DNAs (ssDNA), circular DNAs, and RNAs [12]. The phage genome is packed into a protein capsid, which together form a phage particle. Some phages, such as pleomorphic phages, are further covered by bacterial lipoprotein membrane during budding [13]. As a parasite, the phage life cycle is intertwined with that of the host cell, i.e., bacteria. The phage particle first attaches itself to a host cell by specific recognition of a receptor or other surface moiety of the host; then the phage nucleic acids are inserted into the host cell. Inside the host cell, the phage genome shuts down defense mechanisms of the host and hijacks host cellular components to replicate phage genome and express capsid genes from the phage genome. Eventually, phage genomes and capsids are assembled into progeny phage particles. These phage particles emerge from the host cell, which usually results in cell lysis by phage proteins (**Figure 1**).

Phages were discovered by Twort in 1915 and d'Hérelle in 1917 [12], respectively. Now, it is clear that phages are the most abundant organisms on the earth; they have been found in every environment with bacteria. An estimated 10^{31} phage particles exist on earth. Since the 1940s, phages have been model organisms and have contributed to molecular biology substantially. The most remarkable contributions include revealing the random nature of mutation, the discovery of DNA as the genetic material, and the understanding of gene expression control. With advances in biotechnology, phage display was established by Smith in 1985.

In phage display, the most widely used phage is the M13 strain of filamentous bacteriophage. This type of phage infects F plasmid-containing gram-negative bacteria, such as *Escherichia coli*. Besides M13, other members of filamentous phages such as f1 and fd strains have also been used for phage display [14]. These phages have circular single-stranded DNA (ssDNA). The genome is composed of 11 genes [15], which are classified into three groups by functions. The first group comprises

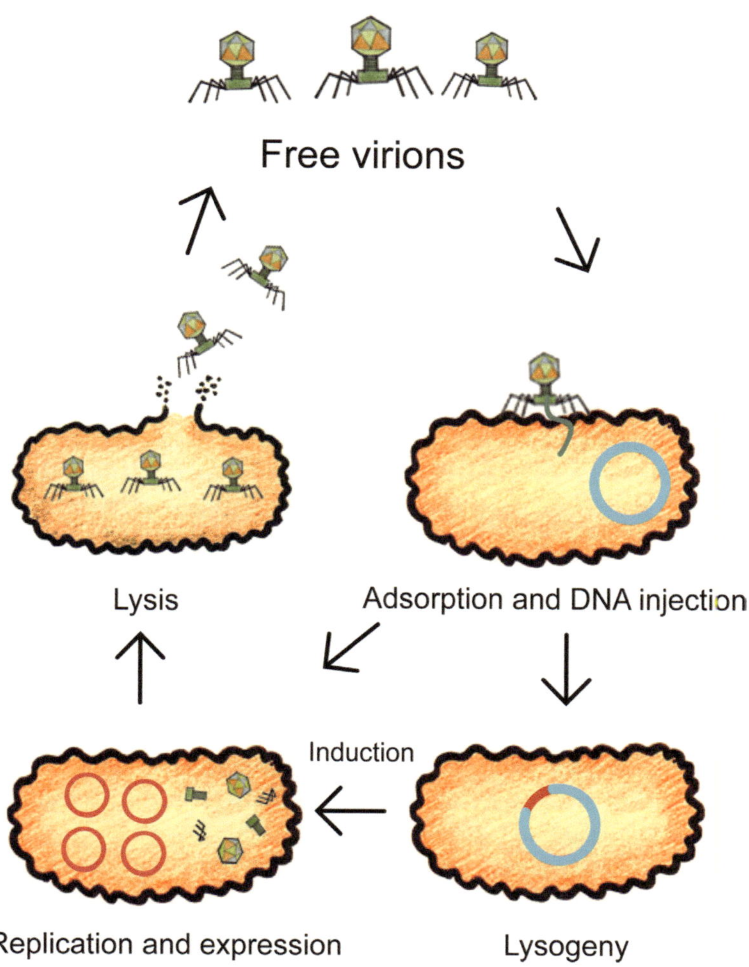

Figure 1.
A scheme presents lytic and lysogenic stages of phage.

capsid proteins: protein III (pIII), pVI, pVII, pVIII, and pIX; the second group is for DNA replication, pII, pV, and pX; the last group consists of proteins for assembly, pI, pIV, and pXI. This genome is contained in a protein coat to form a phage particle of 6.5 nm in width and ~900 nm in length. Normally, filamentous phage is not lytic; thus phages are released from bacteria without bacterial lysis. Instead of phage M13, phage display can also use phagemid, which is simply a plasmid with a phage origin of replication so that the plasmid can be replicated and packaged into phage particles. The phagemid as a cloning vector needs helper phage to complete its infection process and virion packaging [16].

2.2 Phage selection screening methods

2.2.1 Peptide library

Peptide library construction is the first step in selecting peptides with high affinities to the target of interest. Each of the 20 amino acids is encoded by codons, and each codon consists of three consecutive nucleotides. There are four types of nucleotides, denoted as A (adenine), G (guanine), C (cytosine), and T (thymine). A random peptide is constructed by synthesizing an oligonucleotide containing $(NNK)_n$, where N stands for any of the four nucleotides, K stands for either G or T, and n indicates the desired length of the peptide. Note that only G or T is introduced to the third position of a codon because this reduces the frequency of stop codons (NNN generates three stop codons: TAA, TGA and TAG, whereas NNK generates only one codon: TAG). To add a site with N, simply provide an equimolar mixture of A, T, G, and C, and randomly one of them is added at the end of the nucleotide chain. As for a site with K, just provide a mixture of G and T to the reaction [17]. In this way, numerous oligonucleotides are synthesized in parallel, and each oligonucleotide encodes a random peptide. Note that the NNK codon pattern is generated by controlling nucleotide types to be added in reactions. This simple rule can be modified to create particular codons, e.g., allowing no stop codon or creating codons of charged residues.

Peptide libraries often have short peptide lengths, approximately less than 50 residues. The optimum length required for the randomized displayed peptide is often unknown before selection and varies with many factors including the folding properties of the displayed peptide and the target of interest. For a library of random peptides with seven residues, the maximum number of different peptides is 20^7. However, this number is usually unapproachable due to codon degeneracy and early stop codon. In other words, a library usually contains redundant peptides and peptides shorter than desired. On the other hand, the capacity of selection is limited by transfection, i.e., only 10^8–10^{10} phages, each encloses one peptide, can be transformed into *E. coli* by electroporation or other techniques. Taken together, the diversity of a library is important for the success of selection and screening for high affinity peptides. A typical commercially available library archives peptide diversity at the level of 10^9 [4].

In M13 phage, the oligonucleotides encoding random peptides are mostly fused to the N-terminus of pIII, with a spacer as a linker to generate a phage library. Another widely used gene for peptide display is the one encoding pVIII. Both pIII and pVIII are the major and minor capsid proteins, accessible from the outer surface of the phage. pIII has 406 residues, and for each phage, there are in total 3–5 pIII proteins which form a knob-like structure at one end of the phage. This structure is responsible for infection of bacteria via the F-pilus, virion stability, and assembly termination. The peptides linked to pIII for display almost have no restriction on length, facilitating pIII as the mostly targeted for peptide display applications. As for pVIII, it is a short helical protein (50 residues) [18], and about 2700 pVIII molecules are present on the capsid. However, only the three residues at the N-terminus are accessible from the outer surface of the phage. Unlike pIII, pVIII can tolerate only short peptides with less than ten residues to be linked and successfully displayed, which is likely due to interrupted assembly by long peptides. However, this problem can be alleviated by reducing the density of pVIII mutants [19]. Interestingly, some other capsid proteins such as pV1, pVII, and pIX have poorer accessibilities on the phage surface but nevertheless successfully display peptides for screening [19].

2.2.2 In vitro screening

In a phage library, each of the trillion phages displays a single variant of random peptides on its surface. Many methods have been developed to identify the peptides that have high affinities to the target of interest, e.g., a biomarker protein. To this end, biopanning is the most commonly used method, and it has many variations to improve the performance of selection and screening or to accommodate for special targets [20].

Biopanning consists of four basic steps: (i) target immobilization, (ii) phage binding, (iii) washing, and (iv) phage elution. In the first step, the target of interest is purified and immobilized on plates. Some targets are not able to maintain structural integrity after being separated from cells, e.g., some transmembrane proteins. Therefore, whole cells or engineered cells may instead be immobilized on plates or suspended in solutions in this step [21]. In phage binding, the phages in a library are incubated with the plate, and appropriate buffers are also added to facilitate the binding reactions between the displayed peptides on phages and the targets on plates. After that, the plates are washed to remove unbound phages. Phage elution is to acquire the bound phages by disrupting the interaction between the peptides and the targets. This disruption is conducted by changing pH or adding competing ligands, denaturant, or protease. The eluted phages are amplified by infecting *E. coli*. The phages may go through another round of biopanning or are subject to DNA sequencing to determine the peptide sequences (**Figure 2**). These peptides are considered to possess high affinities to the target.

Following several rounds of biopanning, the selected peptides have high affinities to the target but may not have high specificities, i.e., the peptides may also bind to nontargets with high affinities. To this end, subtractive screening can be added to the basic biopanning steps. The subtractive strategy allows phages to interact with nontargets, and thus the unbound phages are candidates that are specific to the target of interest. For example, to identify the peptides that specifically bind to esophageal cancer cells, Zhang et al. use normal human esophageal epithelial cells as the nontarget to perform subtractive screening, followed by screening against esophageal cancer cells. They repeated this procedure for three rounds and discovered two peptides that exhibited higher binding affinities and specificities for the cancer cells, which were validated by enzyme-linked immunosorbent assays (ELISAs), immunofluorescence assays, and immunohistochemistry assays [6].

Improvement of peptide library screening could be performed through small modifications. A stepwise reduction of elution buffer pH in the final round of biopanning reduces low affinity phages, thus effectively further enriching for high affinity phages. Optimized commercial kits are also available, for example, solid-phase screening, solution-sorting screening, kinetic antibody binding screening, and capture-sandwich ELISA screening [22, 23].

2.2.3 In vivo screening

In vivo screening of phage display is designed to isolate tissue-specific binders in living animals. Unlike in vitro screening, this in vivo approach considers the

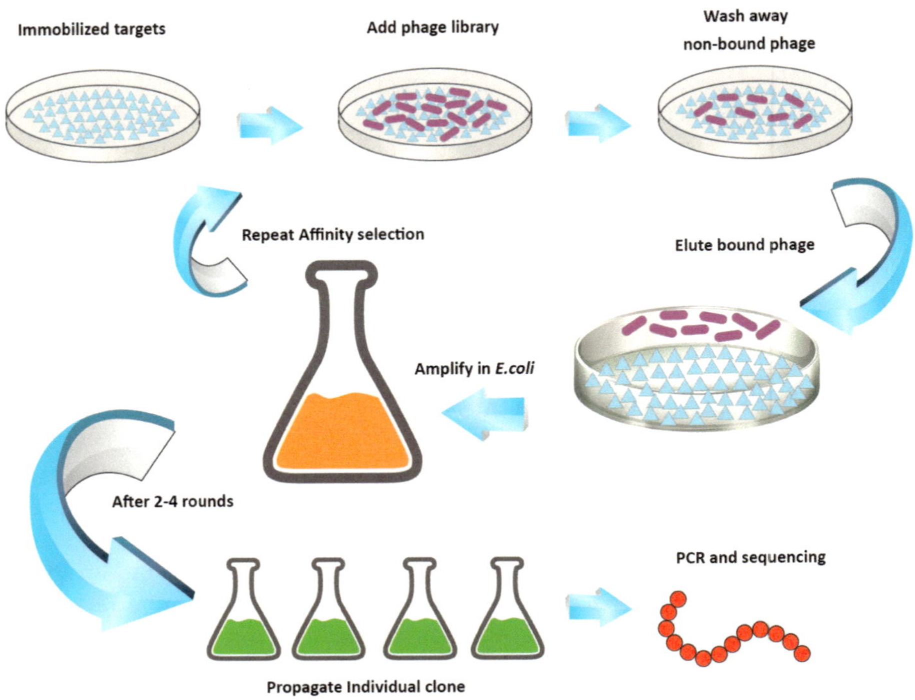

Figure 2.
General scheme of affinity selection of target-specific peptides.

complexity and heterogeneity of the living organism and thus is one step closer to clinical applications. The phage library is administered directly into a living animal and allowed to circulate for a period of time, and then the animal is sacrificed with the desired organ extracted and homogenized in saline. The lysates or pelleted cells of the organ are used to infect *E. coli* so that the phages with high-affinity binding peptides are amplified. This procedure is repeated several rounds, and the resultant phages are sequenced to determine the binding peptide sequences. Similar to in vitro screening, in vivo screening can have a step to wash away unbound phages. It is achieved usually by perfusion of the left ventricle of the animal with saline [24].

In vivo screening faces a more complicated environment than in vitro screening and thus requires extra considerations. Filamentous bacteriophages are often used for in vitro screening; however these viral particles are quite long (>500 nm) and thus may have problems during extravasation to some tissues. For example, M13 phage cannot be used to target liver parenchymal cells due to the impermeability problem [25]. Instead, T7 bacteriophage is used to identify peptides targeting livers because it is smaller in size. An even more challenging issue for in vivo phages is how to avoid immune surveillance for accurate phage displays. Host immune system, particularly reticuloendothelial system, degrades phages quickly and increases nonspecific uptake in the liver and spleen. As expected, the severity of these phenomena reduces within immune-compromised nude and SCID mouse strains. Glycosylation and succinylation to wild-type M13 phages substantially reduce its half-life in murine bloodstream from 4.5 hours to 18 and 1.5 minutes, respectively [26]. This indicates that phages modified to display peptides should have a similarly short half-life, which is consistent with existing data. Therefore, the optimal time of phage library circulation needs to be determined before the actual in vivo phage display experiments. The time varies with phage modifications and targeted tissues.

Phage library administration is another concern for in vivo screening. Administration approaches determine circulation routes of phages in living animals.

The most widely used approach is intravenous delivery. It enables rapid exposure of the phages to vascular receptors of any organ or tissue. However, this approach is inappropriate to the discovery of peptides targeting brain tissues due to the blood-brain barrier. To this end, Wan et al. administered phage library intranasally and identified a peptide targeting the brain, which performs 50-fold better than random control peptides. Other "bloodless" approaches including site-directed phage administrations and transdermal delivery have been used successfully to identify targeting peptides [27].

3. General considerations for phage display to target membrane receptors

Membrane proteins are the most popular targets for diagnostic and therapeutic applications. The structures of membrane proteins are generally composed of three parts, i.e., the extracellular, transmembrane, and intracellular domains. The extracellular domains are the primary targets of drug discovery and diagnosis. About 60% of drug targets are membrane proteins [28]. More specifically, membrane proteins are targeted by the 61% monoclonal antibodies approved or under review as therapeutic drugs throughout Europe and the United States. This prevalence is all attributed to the unique properties of membrane proteins. First, they have various important biological functions such as in signaling and cell channels; thus drugs targeting them can manipulate cellular functions effectively. Second, membrane proteins are presented on cell surface and thus are more accessible than cellular proteins. Some membrane proteins can define cell types. For example, various cluster of differentiation (CD) markers are membrane proteins and define immune cell types. More importantly, diseases usually alter expression levels of membrane proteins. A well-known example is the overexpression of human epidermal growth factor receptor 2 (HER2) membrane protein in 20-25% of breast cancers.

Despite being popular targets, membrane proteins present a limitation for phage display. The key to a successful phage display is the presentation of correctly folded targets, i.e., native structures of extracellular domains (ectodomains) in membrane proteins. This is fairly easy for soluble proteins because they fold correctly in solution while being immobilized on a surface, whereas membrane proteins usually fail to fold into their native structure in solution without membranes [29]. For example, G protein-coupled receptors (GPCRs) constitute the largest class of drug targets but have had limited success in phage display due to their hydrophobic regions and complicated ectodomains, which are usually comprised of the N-terminal chain, parts of the seven transmembrane helices, and their connecting loops. To stabilize the membrane proteins, several methods have been developed. The principle of this method is to engineer the membrane proteins by mutations or adding hydrophilic domains so that they maintain stability in solution for phage display. Note that such engineering must not interfere with the native fold of the ectodomain of membrane proteins.

Another way to circumvent the limitation of unstable membrane proteins in solution is directly biopanning on whole cells instead of immobilizing membrane proteins on a surface as targets for phage display screening. Although this whole cell panning provides membrane proteins in native state as screening targets, it causes extra difficulties. First, the membrane protein may be present at low density on the cell surface; thus other nontargeted membrane proteins generate a high background noise. Second, other components on the surface of cells, such as sugar or lipid polymers, may sequester phage particles for nonspecific uptake into a cell. Taken together, both of these possibilities facilitate the selection of nonspecific ligands, rendering phage display inefficient. To address these newly introduced problems, Jones et al. have reported a modified phage display screening for antibodies. The key innovation is that the cell is transfected so that the target membrane protein and green fluorescent protein (GFP) are highly expressed simultaneously. For phage display, the green fluorescence provides a means to select only the cells with the target membrane protein highly expressed on cell surface using fluorescence-activated cell sorting (FACS) [29]. In other steps of phage display, small but nontrivial modifications are also adopted to improve or overcome the problems of using whole cells. For example, a low pH wash is found effective in removing phage that is present through nonantibody binding. Details of these many small modifications in phage display protocols are reviewed by Alfaleh et al. [30].

A third category of methods has also been developed to create cell membrane-like microenvironments that preserve the stability and integrity of membrane proteins for phage display. In the first method, the membrane protein is purified in the presence of detergents, which forms micelles after exceeding their critical micellar concentration. These micelles mimic cell membrane and have been used in phage display to identify antibodies to the sodium-citrate cotransporter and the fluoride ion channel. However, detergent micelles themselves are unstable and heterogeneous in size, which may cause membrane proteins to unfold or aggregate. In another method, the membrane protein is inserted into a liposome, which is a bilayer structure that more closely mimics the cell membrane than detergent micelles. Recently, nanodiscs and virus-like particles (VLP) have been developed to mimic host membrane protein. Nanodiscs are macromolecular structures that spontaneously assemble when lipids and apolipoprotein A1 or B are mixed. In general, one nanodisc structure contains a lipid bilayer of ~1000 phospholipids bundled by two apolipoproteins. The diameter of a nanodisc is ~10 nm and thus houses only one membrane protein approximately. Nanodiscs can take in a membrane protein through coupled in vitro transcription-translation. A VLP contains viral capsid proteins, lipids, and membrane proteins on its surface, but not a viral genome. It is much more stable than native cells, micelles, and liposomes and thus can withstand wash buffers with detergents, which decreases nonspecific binding in phage display.

In summary, some simple ectodomains of membrane proteins can be directly immobilized on plastic surfaces and treated as soluble proteins in phage display, whereas some other membrane proteins need to be engineered to increase ectodomain stability and integrity before phage display. Various nanoparticles such as detergent micelle, nanodisc, and VLP have been developed to create membrane-like environments to house membrane proteins. Panning directly on wild-type cells or on engineered cells, e.g., with membrane proteins highly expressed, is another way to represent membrane proteins for phage display. Such methods have to manage high background noise from cells. However, these methods have a unique advantage, i.e., consistent with native binding mechanisms. Many membrane receptors, e.g., cytokine receptors, exert their biological functions through ligand binding and dimerization. The ligand may interact with the monomer first and then mediate dimerization for function or bind to the dimer directly and then carry out its function. These two scenarios may both occur depending on the ligand concentrations [31]. Therefore, in those methods without whole cells, the monomer immobilized in phage display may not mimic the targets in vivo, and thus the ligand selected by phage display may not bind to in vivo receptors with high affinities.

4. Phage display provides potential therapeutic and diagnostic agents

Peptide phage display has played an important role in the development of clinically useful therapeutics and diagnostic agents. Peptide-based therapeutics have attracted a significant level of interest in the drug discovery and development industry. First, phage particles themselves can be used as the therapeutic agent. For example, the M13 bacteriophage was used successfully to treat a bacterial infection by delivering DNA encoding for bactericidal toxin proteins [32].

Second, peptides derived from phage display can be used as therapeutic drugs. In 2017, peptide drug annual market was approximately $300–500 million and is estimated to increase 25% each year [33]. Compared to proteins and antibodies, peptides have numerous advantages, for example, low manufacturing costs, better activity and stability, negligible immunogenicity, and superior organ penetration. A number of peptide drugs developed from phage display technique have been approved or are currently in clinical trials. For example, DX-890, an inhibitor of human neutrophil elastase, with potential application in the treatment of pulmonary diseases such as cystic fibrosis and chronic obstructive pulmonary disease, was originated from phage display [34]. Ecallantide, a highly potent inhibitor of human plasma kallikrein, has been approved by the US Food and Drug Administration for the treatment of acute hereditary angioedema [35].

Filamentous phage has also been used as an immunogenic carrier useful in vaccine development, with high immunogenicity, low production cost, and high stability. In addition, phage can also act as a gene-delivery vehicle. For example, phage can deliver functional genes to mammalian cells through receptor-mediated endocytosis.

Phage-derived peptides that bind protein targets with high affinity and specificity can be used as molecular imaging probes. The classic example is octreotide, an eight amino acid cyclized peptide that binds the somatostatin receptor. ^{111}In-DTPA-octreotide (OctreoScan®) has been used successfully to image somatostatin receptor-positive tumors in humans [36].

5. Applications of targeting peptides derived from phages in clinical imaging

Targeted molecular imaging of disease processes, particularly tumor growth and metastasis, has been a focus of many investigations recently. Molecular imaging probes have assisted in the understanding of fundamental biological processes, disease pathologies, as well as pharmaceutical development. Enormous progress has been made in both discovery of imaging probes and development of imaging instruments. Additionally, optical imaging methods provide many advantages over other imaging modalities that include high sensitivity, the use of nonradioactive materials, and safe detection using readily available instruments at moderate cost. Today, in vivo imaging can be applied at preclinical and clinical settings due to significant improvements in engineering technologies, optical systems, and advanced imaging instruments. These technologies in a combination with cutting-edge optical imaging probes provide noninvasive, real-time imaging at macroscopic and cellular levels. Indeed, the combination of numerous NIR probes and targeted ligands, such as antibodies, aptamers, and engineered peptides, has significantly enhanced the performance of optical imaging systems. Recent progress in clinical imaging and the utilization of phage-derived targeting peptides are reviewed below.

5.1 Magnetic resonance imaging

MRI is a medical imaging technique used in radiology to form pictures of the anatomy and the physiological processes of the body, including in healthy and diseased tissues and organs. MRI scanners use strong magnetic fields, electric field gradients, and radio waves to generate images of the organs in the body. After a radiofrequency pulse, MRI detects the relaxation times of magnetic dipoles, such as hydrogen atoms in water and organic compounds, and generates MR signals. MRI offers spatial resolution on the millimeter scale with simultaneous physiological and anatomical correlation. However, MRI requires long scan and postprocessing times and has relatively low sensitivity, thus requiring high doses of magnetic contrast agents.

In a reparative or reactive process, excess fibrous connective tissue (type I collagen) or fibrosis could be formed in a tissue. This is a common consequence in many chronic heart, kidney, liver, lung, or vasculature diseases. The high levels of collagen in fibrosis make it an attractive MRI target. Therefore, a collagen-specific MRI contrast agent was developed. A type I collagen-specific peptide was identified using phage display and subsequently modified to improve affinity for collagen. Conjugate EP-3533 consists of a peptide of 16 amino acids, with 3 amino acids flanking a cyclic peptide of 10 amino acids that is formed through a disulfide bond. The peptide was modified with biphenylalanine and gadolinium to improve collagen binding and sensitivity. EP-3533 was evaluated in a mouse model of aged myocardial infarction. From MR images, EP-3533 was able to enhance the collagen-rich scar, providing specific, high-contrast images compared to adjacent viable myocardium tissues and blood vessels [37].

Another example of the utilization of peptides from phage with MRI in early detection of colorectal cancer (CRC) was reported. Human gastric mucin (MUC5AC) is secreted in the colonic mucus of cancer patients. MUC5AC is a specific marker of precancerous lesions called aberrant crypt foci. MRI can detect the accumulation of MUC5AC in xenograft and mouse stomach. To enhance MRI visualization, peptides that specifically bound MUC5AC were developed using an M13 phage library. Once, the peptide binding MUC5AC (C-PSIYPLL-C, 60C) was identified; it was synthesized and conjugated to biotin and finally to ultra-small particles of iron oxide (USPIOs). The ability of USPIO-60C to detect MUC5AC in vivo was investigated on two xenograft mouse models. A heterogeneous but significant negative enhancement was observed in MUC5AC-secreting tumor postcontrast images 1 hour after intravenous USPIO-60C administration [38]. The results provided in this study supply a proof of concept that targeted contrast agents can be used to detect pathologies earlier than allowed by conventional MRI approaches or clinical assessment.

5.2 Positron emission tomography/single-photon emission computed tomography

PET and SPECT are nuclear medicine tomographic imaging techniques using gamma (γ) rays. Both techniques require the injection of a radioactive tracer. There are three main tracers used in SPECT imaging: technetium-99m, iodine-123, and iodine-131. The radioactive tracer emits gamma rays from the patient. PET and SPECT record high- and low-energy γ-rays emitted from within the body. These imaging modalities have very high sensitivity but relatively low spatial resolution.

VRPMPLQ (VQ) is a heptapeptide sequence first identified by Hsiung and colleagues by screening phage display peptide libraries against fresh human colonic adenomas [39]. In a later study, Shi and team synthesized and evaluated ^{99m}Tc-HYNIC-VQ (HYNIC 5,6-hydrazinonicotinamide) as a SPECT radiotracer for tumor imaging in five different xenograft mouse models (HT-29 human colon cancer, CL187 human colon cancer, BGC823 human gastric cancer, U87MG human glioma, and UM-SCC-22B human head and neck cancer). The images were acquired 1 and 2 hours postinjection. The tumors were clearly visualized at 1 hour postinjection with excellent target-to-background (T/B) contrast. The studies demonstrated that ^{99m}Tc-HYNIC-VQ could provide high-contrast images in different tumor models and an inflammation model [40].

In another study, a radioactive probe targeted to a dysplastic lesion in inflammatory bowel disease (IBD) or early CRC was developed. The cyclic peptide c[Cys-Thr-Pro-Ser-Pro-Phe-Ser-His-Cys]OH (TCP-1) was originally identified in an orthotopic mouse CRC model using phage display selection [41]. TCP-1 peptide was labeled with radioisotope technetium-99 m (^{99m}Tc) and the NIR fluorophore cyanine-7 (Cy7) for molecular imaging. The in vivo images of ^{99m}Tc-TCP-1 in xenografted HCT116 and PC3 prostate cancer models were collected using dynamic or static SPECT. The ^{99m}Tc-TCP-1 or control peptides were administered via intravenous or tail vein injection. Dynamic images of ^{99m}Tc-TCP-1 in HCT116 colon cancer xenograft mice exhibited that the tumor could be detected in 15–30 minutes after injection and remained visible until 180 minutes. The data demonstrated the feasibility of TCP-1-targeted detection of colorectal tumor [42].

5.3 Photoacoustic tomography

Photoacoustic tomography (PAT) is an emerging imaging technique that demonstrates great potential for preclinical research and clinical applications. PAT is a hybrid system that detects the acoustic energy of endogenous chromophore or exogenous contrast agent optical absorption. PAT generates high-resolution images in both the optical ballistic and diffusive regimes due to less ultrasound scatter in tissue. Over the past decade, the photoacoustic technique has been developing rapidly, leading to exciting findings and applications.

Epidermal growth factor receptor (EGFR) is highly overexpressed in hepatocellular carcinoma (HCC). Therefore, it is a potential cell surface molecule for in vivo targeted imaging of HCC. A peptide specific for EGFR previously reported by Zhou and team was conjugated to Cy5.5 dye. Nude mice were injected with EGFR overexpressed human HCC cells. A 2D ultrasound scanner and MRI system were used to monitor tumor growth in the mice. Cy5.5-labeled EGFR and control peptides were injected to the mice separately. Photoacoustic images were recorded periodically for 24 hours. At 3 hours postinjection, the maximum photoacoustic signal in tumors was seen and results in high-contrast images of tumors beneath the skin. The T/B ratio was significantly different between the EGFR and control peptide. The signal was diminished by 24 hours [43]. From the data, a peptide specific for EGFR can detect HCC xenograft tumors in vivo with photoacoustic imaging.

5.4 Optical endomicroscopy

Optical imaging offers several unique advantages. Optics is nonionizing and provides resolution on the micron scale. Another advantage of optical imaging is the ability to collect images in real time in comparison to other imaging modalities, such as MRI and PET.

Endoscopes are thin, flexible instruments that provide a macroscopic view of the large mucosal surfaces in hollow organs internal to the human body. Endomicroscopy employs high numerical aperture (NA) optics to provide a small field-of-view (FOV) with micron-level resolution for observing subcellular features. It commonly requires scaling down the size of a conventional microscope design into a miniature package. Novel optical designs and scanning mechanisms have been developed to improve imaging performance for both endoscopy and endomicroscopy. These instruments provide a unique opportunity for early cancer detection and prevention by allowing biopsy or resection to be performed concurrently with diagnosis.

5.4.1 Wide-field fluorescence endoscopy

Previously, all CRCs were believed to arise from adenomas that progress through the traditional adenocarcinoma sequence. Recently, this pathway has been found to account for approximately 60% of CRCs, and up to 35% are now attributed to the serrated pathway [44]. White-light endoscopy (WLE) that is normally used in colonoscopy is sensitive to gross morphologic abnormalities, such as polyps. Dysplasia that is flat in morphology, focal in size, and patchy in distribution appears "invisible" on conventional wide-field endoscopy. Therefore, imaging methods with improved contrast and sensitivity to molecular rather than morphological properties that could improve early detection and prevention of CRC are in needed. There are several reports on discovery and validation of targeted peptides derived from phage display for early detection of CRC [45].

One group used phage display to identify a peptide that binds to dysplastic colonic mucosa in vivo in a genetically engineered mouse model of colorectal tumorigenesis, *CPC;Apc* [46, 47]. A peptide, QPIHPNNM, was isolated after several rounds of in vivo T7 library biopanning. The peptide was synthesized, fluorescently labeled, and purified. The peptide was sprayed topically in mouse distal colon. The wide-field fluorescent videos were recorded. After quantitative image analysis, the fluorescent-labeled peptide was significantly bound twofold greater to the colonic adenomas when compared to the control peptide. The target peptide also showed minimal binding to an activated Kras^{G12D} mutant mouse model that demonstrates hyperplastic polyp-like features used as a control hyperplastic model that does not progress to carcinoma (**Figure 3A**) and the lumen of a *CPC;Apc* bred mouse negative (**Figure 3B**).

c-Met overexpression has been shown to occur as an early event in colorectal adenocarcinoma. A peptide that has high affinity for the extracellular domain of human c-Met was discovered using an M-13 phage display library. GE-137 is a water-soluble cyclic peptide (AGSCYCSGPPRFECWCYETEGT) labeled with a cyanine dye with a high affinity for human c-Met. The quantitative biodistribution, pharmacokinetics, binding specificity, and qualitative fluorescence of GE-137 were assessed in a CRC xenograft mouse model using subcutaneous injection of the c-Met-expressing human CRC cell line HT29. Intravenously administered GE-137 accumulated in the c-Met-expressing tumor xenografts and left a fluorescent signal in the tumors and kidneys 120 and 240 minutes

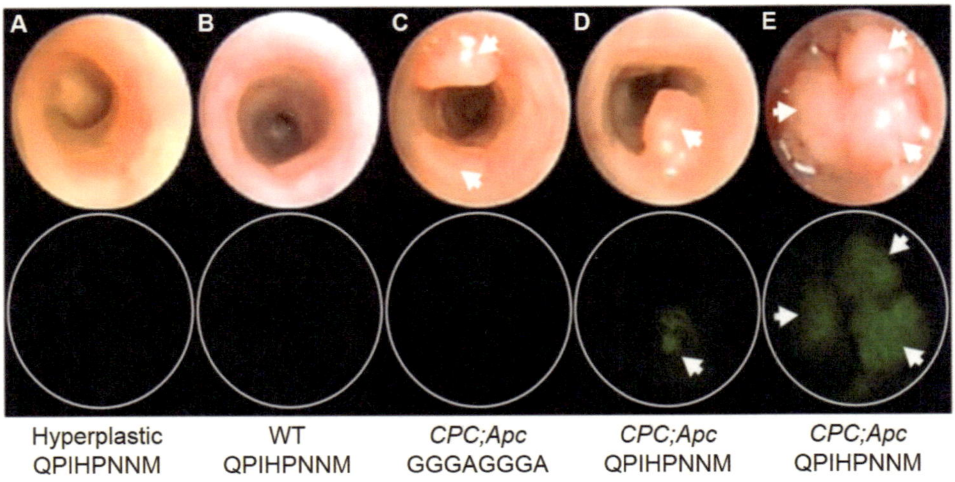

Figure 3.
Images from wide-field endoscopy videos after topical application of fluorescence-labeled peptides. The top and bottom panels represent frames from white light and fluorescence, respectively. (A) The hyperplastic epithelium after QPI peptide application, (B) the lumen of a CPC;Apc bred mouse negative for Cre recombinase (control litter mate), (C) single adenoma after control peptide application, (D) single adenoma, and (E) multiple adenomas in a CPC;Apc mouse after QPI peptide application. Used with permission [47].

after injection (**Figure 4**). In a pilot study in 15 patients at high risk of colorectal neoplasia, a total of 101 lesions were detected during first inspection with white light (WL), and an additional 22 were detected during second inspection with dual WL/fluorescence (FL). After immunohistochemical analysis, 36 hyperplastic lesions and 8 serrated polyps were identified. Most of these were visible in fluorescent mode (94 and 100%, respectively), and the majority (78 and 87%, respectively) showed increased fluorescence. From the data, GE-137 peptide showed some specificity to hyperplasia and serrated lesions in CRC mouse model and patients [48].

Overexpression of EGFR has been reported in as high as 97% of colonic adenocarcinomas, and it is a validated biomarker for CRC. In one study, the extracellular domain of EGFR (EGFR-ECD) was expressed and purified [49]. A library of M13 bacteriophage was used to select peptide candidates that bind specifically to EGFR-ECD. A peptide, QRHKPRE, that is specific for EGFR was developed and validated. Peptide binding to cells occurred within 2.46 minutes and had an affinity of 50 nM. A NIR fluorescence endoscope was used to perform *in vivo* imaging to validate peptide binding to spontaneous colonic adenomas in a *CPC;Apc* mouse model via topical administration (**Figure 5**). T/B ratios of polyps and flat lesions were 4.0 ± 1.7 and 2.7 ± 0.7, respectively. Subsequently, specific peptide binding to human colonic adenomas was assessed on immunohistochemistry and immunofluorescence. On human colonic specimens, greater intensity from peptide binding to dysplasia than normal was found with a 19.4-fold difference.

In another study, a phage-derived peptide was tested for specific binding to sessile serrated adenomas (SSAs) in proximal colon which accounts for 35% in CRC. Joshi and team used phage display to identify a peptide that binds specifically to SSAs. Many SSA cells have the V600E mutation in BRAF. Therefore, peptide selection was performed with an M13 Ph.D.-7 phage display library using a biopanning strategy with subtractive hybridization with HT29 colorectal cancer cells containing the V600E mutation in BRAF [45]. Binding of fluorescently labeled peptide, KCCFPAQ, to colorectal cancer cells was evaluated with confocal fluorescence

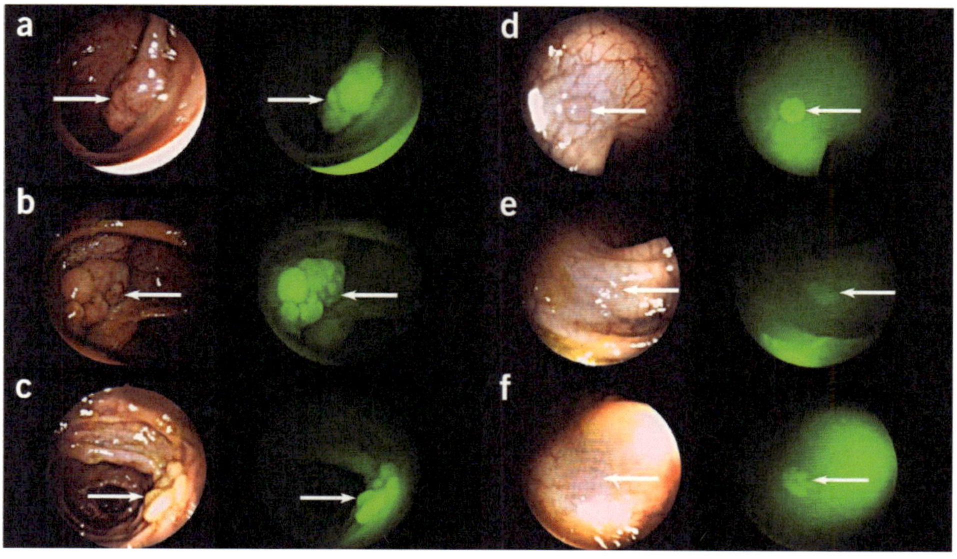

Figure 4.
WL and FL images of representative lesions. (a–c) The lesions that are visible in WL show increased fluorescence. (d) A lesion that is visible in WL has enhanced visibility in FL. (e, f) Flat lesions that were only visible in FL. Used with permission [48].

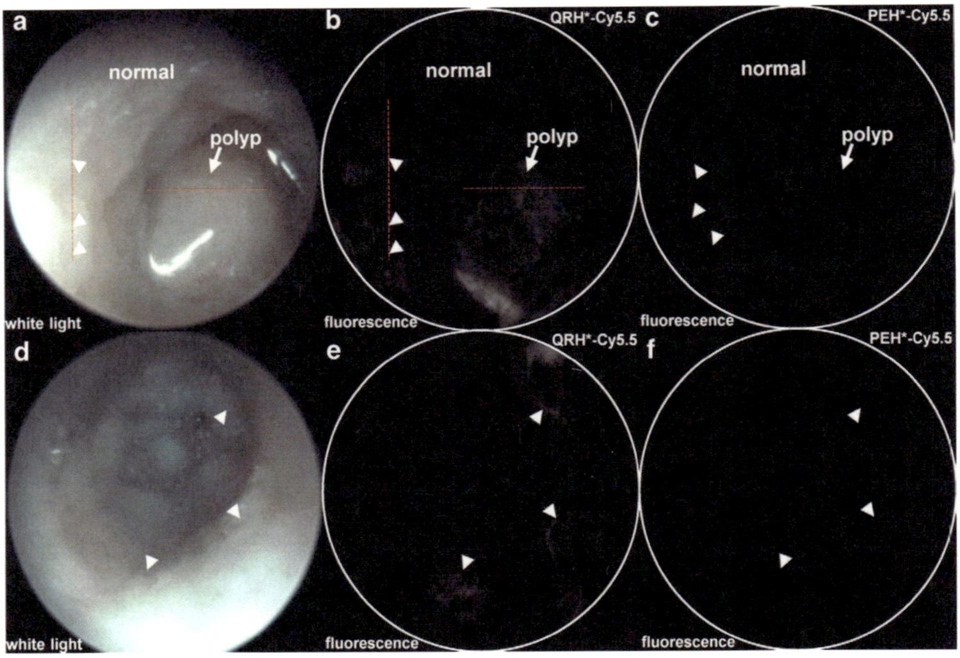

Figure 5.
In vivo imaging of colon in CPC;Apc mouse. *(a) WL image of colon in CPC;Apc mouse shows the presence of polyp (arrow). (b) NIR fluorescence image after topical administration of QRH*-Cy5.5 shows increased intensity from polyp (arrow) and several flat lesions (arrowheads). (c) Image with Cy5.5-control peptide (PEH*-Cy5.5) shows minimal signal. (d) WL image shows no grossly visible lesions (polyps). (e) NIR fluorescence image with QRH*-Cy5.5 shows the presence of flat lesions (arrowheads). (f) Image with Cy5.5-control peptide shows minimal signal. Used with permission [49].*

microscopy. The peptide showed an apparent dissociation constant of K_d = 72 nM and an apparent association time constant of K = 0.174/minute. Toxicity was also assessed in rats. In the clinical safety study, fluorescently labeled peptide was topically administered, using a spray catheter, to the proximal colon of 25 subjects undergoing routine outpatient colonoscopies. Subsequently, endoscopists resected identified lesions, which were analyzed histologically by gastrointestinal pathologists. Fluorescence intensities of SSAs were compared with those of normal colonic mucosa. During fluorescence imaging of patients during endoscopy, regions of SSA had 2.43-fold higher mean fluorescence intensity than that for normal colonic mucosa. Fluorescence labeling distinguished SSAs from normal colonic mucosa with 89% sensitivity and 92% specificity.

5.4.2 Confocal laser endomicroscopy

Confocal laser endomicroscopy (CLE) is an endoscopic modality that based on tissue illumination using a low power laser and the subsequent detection of fluorescent light that is reflected back from the tissue through a pinhole. The term "confocal" refers to the alignment of both illumination and collection systems in the same plane. This alignment dramatically increases the spatial resolution of CLE and enables cellular imaging and evaluation of tissue architecture at the focal plane. CLE can be used to guide biopsies and has been demonstrated in a number of clinical studies to detect cancer in the digestive tract, bladder, cervix, ovary, oral cavity, and lungs. CLE requires the use of a fluorescent contrast agent to enhance visualization of cells. Contrast agents can be administered intravenously or topically. Intravenous fluorescein sodium and acriflavine are widely used contrast agents.

CLE can be performed shortly following injection with its fluorescence lasting approximately 30 minutes. However, due to the lack of specificity of conventional contrast agents, there is an increasing use of tissue-specific binding molecular probes in CLE. The studies that used CLE in combination with a targeting peptide derived from phage display are summarized below.

Hsiung et al. used an M13 phage library to identify peptides that would specifically bind dysplastic colonic mucosa from fresh human colonic biopsies [39]. A peptide (VRPMPLQ) that bound to colonic dysplasia was identified, synthesized, and conjugated with fluorescein for in vivo testing in a pilot study in patients undergoing routine colonoscopy using a flexible-fibered confocal microscope. Fluorescence images and videos of bound topically administered peptide collected in vivo showed that the selected peptide bound more strongly to dysplastic colonocytes than to the adjacent normal mucosa in the same subject with 81% sensitivity and 82% specificity.

In another study, Qiu and team demonstrated vertical cross-sectional (XZ-plane) images of NIR fluorescence with a handheld dual-axis confocal endomicroscope that revealed a specific binding of a Cy5.5-labeled peptide (LTTHYKLGGGSK-Cy5.5) to premalignant colonic mucosa in mice [50]. This targeting peptide was selected using in vivo phage display technology in a *CPC;Apc* mouse model which developed adenomas spontaneously in the distal colon. NIR vertical cross-sectional fluorescence images of fresh mouse colonic mucosa demonstrate histology-like imaging performance as shown in **Figure 6**. The peptide showed specific binding and distinguished premalignant colonic mucosa from normal mucosa.

Gastric cancer vessels may have many differentiating characteristics compared to normal vessels. However, identification of gastric cancer vascular endothelial cells is difficult. Co-culture of gastric cancer cells and vascular endothelial cells was suggested to simulate gastric solid tumor mass. In a study, GEBP11, a nine amino

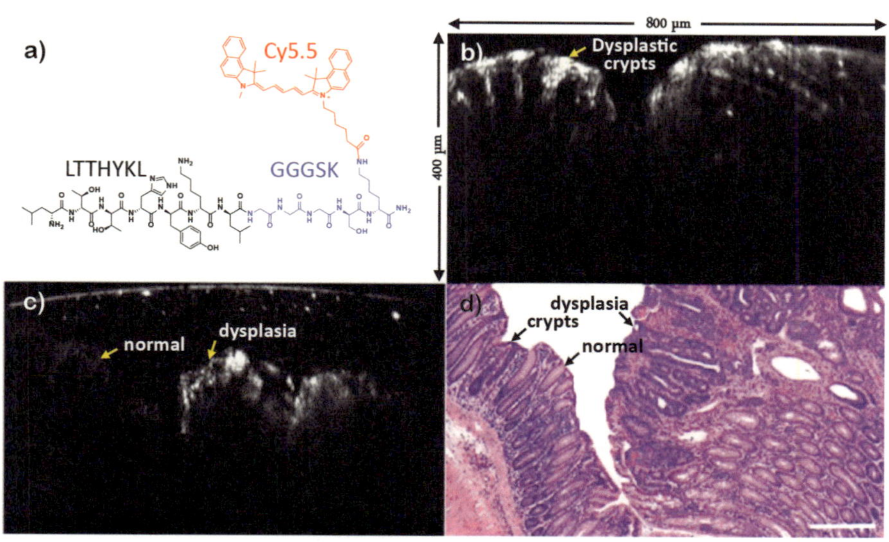

Figure 6.
Vertical cross-sectional image of colonic dysplasia. (a) Chemical structure of LTTHYKL peptide with GGGSK linker and Cy5.5 fluorophore. (b) NIR fluorescence image from CPC; Apc mouse colon ex vivo shows vertically oriented dysplastic crypts. (c) The border between normal colonic mucosa and dysplasia shows increased contrast from specific binding of the LTT-Cy5.5 peptide. (d) Corresponding histology (H&E), scale bar 200 µm. Used with permission [50].*

acid vascular homing peptide, was screened and identified using Ph.D.C7C phage display peptide library kit panning against Co-HUVECS cells [51]. Liu et al. used FITC-GEBP11 to identify gastric cancer in mouse model [52]. A whole-body fluorescent imaging was first used to screen for the strongest specific fluorescent signal in xenograft models after tail vein injection of FITC-GEBP11. A specific signal was observed in both subcutaneous and orthotopic xenograft models in vivo, whereas the group injected with FITC-URP, a control peptide, showed no fluorescent signals. In addition, neoplastic and nonneoplastic gastric mucosae obtained from the patients were incubated with FITC-GEBP11 or FITC-URP for 30 minutes and were scanned with CLE. A specific signal of GEBP11 was observed in 26/28 neoplastic human specimens and in 8/28 samples of nonneoplastic specimens ($p < 0.01$).

Sturm and team developed a peptide (ASYNYDA) that binds specifically to high-grade esophageal dysplasia and adenocarcinoma using phage display technology [53]. After peptide specific binding validation in human esophageal cancer specimens, they applied peptide topically and performed confocal endomicroscopy in 25 patients. The targeting peptide showed 3.8-fold greater fluorescence intensity for esophageal neoplasia compared with Barrett's esophagus and squamous epithelium with high sensitivity and specificity. The peptide revealed no toxicity in animals or patients.

In a pilot study, Palma and team identified dysplasia lesions in ulcerative colitis (UC) patients using CLE and a fluorescently labeled peptide [54]. A phage-derived peptide (VRPMPLQ) was synthesized and conjugated with fluorescein. Eleven suspected dysplasia specimens were collected from nine UC patients. Specimens were stained with the peptide and subsequently inspected by CLE. The CLE images were correlated to histological results from specialists. The peptide showed a different pattern on dysplastic mucosa compared to nondysplastic lesions. However, due to several restrictions of this study, further studies on larger UC patients are required for systematic validation.

ErbB2 expression in early breast cancer can predict tumor aggressiveness and clinical outcomes in patients. Up to 30% of all breast cancers express ErbB2, also known as HER2. Immunohistochemistry is commonly used to evaluate ErbB2 expression, but it has limitations due to tumor heterogeneity. Therefore, the use of a specific biomarker for ErbB2 is increasingly popular. One study used a NIR-labeled ErbB2 peptide and a handheld dual-axis CLE to detect in human xenograft breast tumors and human specimens [55]. As a result, they found significantly greater peptide binding to xenograft breast cancer in vivo and to human specimens of invasive ductal carcinoma that express ErbB2 ex vivo. From the data, a miniature dual-axis confocal fluorescence endomicroscope with ErbB2-specific peptide could be implemented to support future image-guided surgery.

5.5 Multimodality imaging

Radiation-induced pulmonary fibrosis (RIPF) is a serious side effect of radiation therapy, especially in lung and breast cancers. Computed tomography (CT) imaging is currently utilized to identify and monitor RIPF. However, anatomical change interference is a major limitation of CT. Therefore, RIPF detection and observing techniques need to be improved. Collagen accumulation is common in fibrosis. In one study, a collagen-targeting peptide was fabricated to maximize the visualization of fibrosis using fluorescence endomicroscope imaging [37]. The probe showed moderate binding ability to collagen in a fibrosis in vitro binding assay and on lung tissue specimens. The probe showed a similar binding pattern on lung specimens compared to antibody. But its sensitivity was not as good as the collagen-binding antibody.

In another study, Zhang and team evaluated the potential applicability of GEBP11 peptides in gastric cancer diagnosis and radiotherapy. They developed iodine 131-labeled GEBP11 peptides and derivatives [56]. The clinical potential of GEBP11 peptides was determined with multimodality imaging methods. Cerenkov and SPECT imaging showed significantly higher tumor uptake for ^{131}I-2PEG-(GEBP11)$_3$ trimer compared to monomer. Higher tumor accumulation and better T/B ratio of ^{131}I-2PEG-(GEBP11)$_3$ trimer were observed. Treating with ^{131}I-2PEG-(GEBP11)$_3$ trimer exhibited a significant tumor growth suppression compared to control and monomer groups. The tumor volume and vasculature decreased significantly after treatment with ^{131}I-2PEG-(GEBP11)$_3$ trimer, resulting in prolonged survival time. In addition, ^{131}I-2PEG-(GEBP11)$_3$ trimer showed no significant hepatic or renal toxicity. In conclusion, ^{131}I-2PEG-(GEBP11)$_3$ trimer could be a potential ligand used to identify gastric cancer and in antiangiogenic therapy.

6. Future prospects

The discovery and development of therapeutic drugs and diagnostic probes are a time-consuming, expensive, and complex process. The processes involve experts from a wide range of disciplines such as medicinal chemistry, biochemistry, molecular biology, medicine, and pharmacology. It has been estimated that from about 10,000 new chemical entities identified, only one will reach the market in an average time of 16 years. Phage display, and particularly peptide phage display, has played a major role in the development pipeline for bringing peptide therapeutics into the clinic. Phage-derived peptides play an important role in disease detection and therapy, including in clinical imaging. The potential of peptides in preclinical and clinical molecular imaging is tremendous. Molecular imaging offers invaluable opportunities to explore complex disease-related biological processes at the molecular level in vivo. The emergence of current molecular imaging technologies is dependent not only on the progress of imaging systems but, more importantly, also on molecular imaging probes. Peptide-based imaging has now become an established approach in nuclear imaging, and its application is expanding to other imaging modalities. Considering the emergence of novel library designs and innovative selection strategies, we are confident that phage-derived peptides will continue to be promising biomarkers for early cancer detection, in metabolic abnormalities, and in personalized medicine.

Author details

Supang Khondee[1] and Wibool Piyawattanametha[2,3]*

1 School of Pharmaceutical Sciences, University of Phayao, Phayao, Thailand

2 Faculty of Engineering, Department of Biomedical Engineering, King Mongkut's Institute of Technology Ladkrabang (KMITL), Bangkok, Thailand

3 Institute for Quantitative Health Science and Engineering (IQ), Michigan State University, Michigan, USA

*Address all correspondence to: wibool@gmail.com

References

[1] Lee S, Xie J, Chen X. Peptide-based probes for targeted molecular imaging. Biochemistry. 2010;**49**(7):1364-1376

[2] De Jong M et al. Tumor imaging and therapy using radiolabeled somatostatin analogues. Accounts of Chemical Research. 2009;**42**(7):873-880

[3] Wu AM, Olafsen T. Antibodies for molecular imaging of cancer. The Cancer Journal. 2008;**14**(3):191-197

[4] Derda R et al. Diversity of phage-displayed libraries of peptides during panning and amplification. Molecules. 2011;**16**(2):1776-1803

[5] Smith GP. Filamentous fusion phage: Novel expression vectors that display cloned antigens on the virion surface. Science. 1985;**228**(4705):1315-1317

[6] Zhang ZF et al. Screening and selection of peptides specific for esophageal cancer cells from a phage display peptide library. Journal of Cardiothoracic Surgery. 2014;**9**(1):76

[7] Witt H et al. Identification of a rhabdomyosarcoma targeting peptide by phage display with sequence similarities to the tumour lymphatic-homing peptide LyP-1. International Journal of Cancer. 2009;**124**(9):2026-2032

[8] Hui X et al. Specific targeting of the vasculature of gastric cancer by a new tumor-homing peptide CGNSNPKSC. Journal of Controlled Release. 2008;**131**(2):86-93

[9] Wang W et al. Near-infrared optical imaging of integrin $\alpha v\beta 3$ in human tumor xenografts. Molecular Imaging. 2004;**3**(4):15353500200404148

[10] Ye Y et al. Design, synthesis, and evaluation of near infrared fluorescent multimeric RGD peptides for targeting tumors. Journal of Medicinal Chemistry. 2006;**49**(7):2268-2275

[11] Comeau AM et al. Exploring the prokaryotic virosphere. Research in Microbiology. 2008;**159**(5):306-313

[12] Ofir G, Sorek R. Contemporary phage biology: From classic models to new insights. Cell. 2018;**172**(6):1260-1270

[13] Drulis-Kawa Z, Majkowska-Skrobek G, Maciejewska B. Bacteriophages and phage-derived proteins, application approaches. Current Medicinal Chemistry. 2015;**22**(14):1757-1773

[14] Murphy FA et al. Virus Taxonomy: Classification and Nomenclature of Viruses. Vol. 10. New York: Springer Science & Business Media; 2012

[15] Russel M, Lowman HB, Clackson T. Introduction to phage biology and phage display. In: Phage Display: A Practical Approach. New York: Oxford University Press; 2004. pp. 1-26

[16] Barbas CF, et al. Phage Display: A Laboratory Manual. New York: Cold Spring Harbor Laboratory Press; 2001. pp. 1-24

[17] Pal G, Fellouse FA. Methods for the construction of phage-displayed libraries. In: Phage Display in Biotechnology and Drug Discovery. Florida: CRC Press; 2005. pp. 131-162

[18] Zeri AC et al. Structure of the coat protein in fd filamentous bacteriophage particles determined by solid-state NMR spectroscopy. Proceedings of the National Academy of Sciences. 2003;**100**(11):6458-6463

[19] Bratkovic T. Progress in phage display: Evolution of the technique and

its applications. Cellular and Molecular Life Sciences. 2010;**67**(5):749-767

[20] Pande J, Szewczyk MM, Grover AK. Phage display: Concept, innovations, applications and future. Biotechnology Advances. 2010;**28**(6):849-858

[21] Hamzeh-Mivehroud M, Mahmoudpour A, Dastmalchi S. Identification of new peptide ligands for epidermal growth factor receptor using phage display and computationally modeling their mode of binding. Chemical Biology & Drug Design. 2012;**79**(3):246-259

[22] Matz J, Chames P. Phage display and selections on purified antigens. In: Antibody Engineering. Heidelberg: Springer; 2012. pp. 213-224

[23] Smolarek D, Bertrand O, Czerwinski M. Variable fragments of heavy chain antibodies (VHHs): A new magic bullet molecule of medicine? Advances in Hygiene & Experimental Medicine. 2012;**66**:348-358

[24] Kolonin MG et al. Synchronous selection of homing peptides for multiple tissues by in vivo phage display. The FASEB Journal. 2006;**20**(7):979-981

[25] Ludtke JJ et al. In vivo selection and validation of liver-specific ligands using a new T7 phage peptide display system. Drug Delivery. 2007;**14**(6):357-369

[26] Molenaar TJ et al. Uptake and processing of modified bacteriophage M13 in mice: Implications for phage display. Virology. 2002;**293**(1):182-191

[27] Wu M et al. Mapping alveolar binding sites in vivo using phage peptide libraries. Gene Therapy. 2003;**10**(17):1429

[28] Arinaminpathy Y et al. Computational analysis of membrane proteins: The largest class of drug targets. Drug Discovery Today. 2009;**14**(23-24):1130-1135

[29] Jones ML et al. Targeting membrane proteins for antibody discovery using phage display. Scientific Reports. 2016;**6**:26240

[30] Alfaleh MA et al. Strategies for selecting membrane protein-specific antibodies using phage display with cell-based panning. Antibodies. 2017;**6**(3):10

[31] Schooltink H, Rose-John S. Designing cytokine variants by phage-display. Combinatorial Chemistry & High Throughput Screening. 2005;**8**(2):173-179

[32] Westwater C et al. Use of genetically engineered phage to deliver antimicrobial agents to bacteria: An alternative therapy for treatment of bacterial infections. Antimicrobial Agents and Chemotherapy. 2003;**47**(4):1301-1307

[33] Lau JL, Dunn MK. Therapeutic peptides: Historical perspectives, current development trends, and future directions. Bioorganic & Medicinal Chemistry. 2017;**26**:2700-2707

[34] Rothe A, Hosse RJ, Power BE. In vitro display technologies reveal novel biopharmaceutics. The FASEB Journal. 2006;**20**(10):1599-1610

[35] Zuraw, B, Yasothan U, Kirkpatrick P. Ecallantide. Nature Reviews Drug Discovery. 2010;**9**:189-190

[36] Froidevaux S, Eberle AN. Somatostatin analogs and radiopeptides in cancer therapy. Peptide Science. 2002;**66**(3):161-183

[37] Caravan P et al. Collagen-targeted MRI contrast agent for molecular imaging of fibrosis. Angewandte Chemie International Edition. 2007;**46**(43):8171-8173

[38] Rossez Y et al. Early detection of colonic dysplasia by magnetic resonance molecular imaging with a contrast agent raised against the colon cancer marker MUC5AC. Contrast Media & Molecular Imaging. 2016;**11**(3):211-221

[39] Hsiung PL et al. Detection of colonic dysplasia in vivo using a targeted heptapeptide and confocal microendoscopy. Nature Medicine. 2008;**14**(4):454

[40] Shi J et al. Technetium 99m–labeled VQ peptide: A new imaging agent for the early detection of tumors or premalignancies. Molecular Imaging. 2013;**12**(5):7290.2012. 00047

[41] Li ZJ et al. A novel peptide specifically targeting the vasculature of orthotopic colorectal cancer for imaging detection and drug delivery. Journal of Controlled Release. 2010;**148**(3):292-302

[42] Liu Z et al. Characterization of TCP-1 probes for molecular imaging of colon cancer. Journal of Controlled Release. 2016;**239**:223-230

[43] Zhou Q et al. In vivo photoacoustic tomography of EGFR overexpressed in hepatocellular carcinoma mouse xenograft. Photoacoustics. 2016;**4**(2):43-54

[44] Snover DC. Update on the serrated pathway to colorectal carcinoma. Human Pathology. 2011;**42**(1):1-10

[45] Joshi BP et al. Detection of sessile serrated adenomas in the proximal colon using wide-field fluorescence endoscopy. Gastroenterology. 2017;**152**(5):1002-1013. e9

[46] Elahi SF et al. Targeted imaging of colorectal dysplasia in living mice with fluorescence microendoscopy. Biomedical Optics Express. 2011;**2**(4):981-986

[47] Miller SJ et al. In vivo fluorescence-based endoscopic detection of colon dysplasia in the mouse using a novel peptide probe. PLoS One. 2011;**6**(3):e17384

[48] Burggraaf J et al. Detection of colorectal polyps in humans using an intravenously administered fluorescent peptide targeted against c-Met. Nature Medicine. 2015;**21**(8):955

[49] Zhou J et al. EGFR overexpressed in colonic neoplasia can be detected on wide-field endoscopic imaging. Clinical and Translational Gastroenterology. 2015;**6**(7):e101

[50] Qiu Z et al. Targeted vertical cross-sectional imaging with handheld near-infrared dual axes confocal fluorescence endomicroscope. Biomedical Optics Express. 2013;**4**(2):322-330

[51] Liang S et al. Screening and identification of vascular-endothelial-cell-specific binding peptide in gastric cancer. Journal of Molecular Medicine. 2006;**84**(9):764-773

[52] Liu L et al. In vivo molecular imaging of gastric cancer in human-murine xenograft models with confocal laser endomicroscopy using a tumor vascular homing peptide. Cancer Letters. 2015;**356**(2):891-898

[53] Sturm MB et al. Targeted imaging of esophageal neoplasia with a fluorescently labeled peptide: First-in-human results. Science Translational Medicine. 2013;**5**(184):184ra61-184ra61

[54] De Palma GD et al. Detection of colonic dysplasia in patients with ulcerative colitis using a targeted fluorescent peptide and confocal laser endomicroscopy: A pilot study. PLoS One. 2017;**12**(6):e018050e9

[55] Gao Z et al. In vivo near-infrared imaging of ErbB2 expressing

breast tumors with dual-axes confocal endomicroscopy using a targeted peptide. Scientific Reports. 2017;**7**(1):14404

[56] Zhang J et al. Targeted radiotherapy with tumor vascular homing trimeric GEBP11 peptide evaluated by multimodality imaging for gastric cancer. Journal of Controlled Release. 2013;**172**(1):322-329